New Jersey Ferns and Fern-Allies

New Jersey Ferns and Fern-Allies

James D. Montgomery
and
David E. Fairbrothers

Illustrated by Kathleen L. John-Alder

Rutgers University Press
New Brunswick, New Jersey

Library of Congress Cataloging-in-Publication Data

Montgomery, James D., 1937–
New Jersey ferns and fern-allies / James D. Montgomery and David E. Fairbrothers.
p. cm.
Includes bibliographical references (p.) and index.
ISBN 0-8135-1817-2
1. Ferns—New Jersey—Identification. 2. Pteridophyta—New Jersey—Identification.
3. Ferns—New Jersey—Pictorial works. 4. Pteridophyta—New Jersey—Pictorial works.
I. Fairbrothers, David E. II. Title.
QK525.5.N5M66 1992
587′.3′09749—dc20 91-40920
CIP

British Cataloging-in-Publication information available

Manufactured in the United States of America

Contents

Acknowledgments

This book was first suggested to the senior author by Dr. Marion A. Johnson, former dean of the Graduate School and professor of botany at Rutgers University, as an extension of the research being conducted by Rutgers botanists on the flora of New Jersey. Dr. Johnson conducted field studies with M. A. Chrysler and J. L. Edwards and so was familiar with the background leading to the publication of *The Ferns of New Jersey* in 1947. Although this book was published prior to Manton's (1950) publication, which is usually considered the beginning of modern fern taxonomy, it included much advanced information not present in the current manuals of the day (Fernald 1950). This included recognition of the modern species concepts in *Dryopteris* and the recognition of the importance of hybrids in fern identification problems.

Much information was collected by M. Y. Hough and J. S. Meyers, assistant curators of the Chrysler Herbarium at Rutgers. Several individuals assisted us with field studies of New Jersey ferns, including R. C. Cahayla-Wynne, C. W. Haefele, III, D. G. Huttleston, and D. B. Snyder. A number of other botanists have provided taxonomic insight or assistance with particular problematic groups of pteridophytes, including J. M. Beitel, D. B. Lellinger, J. T. Mickel, F. S. Wagner, and W. H. Wagner, Jr. We express our appreciation to these people for their time and assistance. We especially thank W. Carl Taylor for his contribution of *Isoetes,* and our botanical illustrator Kathleen L. John-Alder for her patient and careful work. Grants were received from the McCormick Publication Fund of the New Jersey Academy of Science and the Willowwood Foundation.

This book is dedicated to the natural beauty of the state of New Jersey.

J.D.M.
Berwick, PA

D.E.F.
Piscataway, NJ

January 1992

New Jersey Ferns and Fern-Allies

Historical Introduction

Ferns and fern-allies (pteridophytes) number approximately 10,000 species and have been hunted by amateur and professional botanists for hundreds of years. They were used extensively in the past because of assumed magical or medicinal properties; however, only a few species have proven to be useful for either of these purposes (Lellinger 1985). The Victorian period (1830–1901) produced a mania for fern collecting and cultivation in western Europe and to a lesser extent in America, as people attempted to discover new and exotic variants for cultivation. This craze faded with the turn of the twentieth century.

Since the early 1970s there has been a rejuvenation of enthusiasm among gardeners and naturalists for growing ferns in both large and small spaces. Pteridophytes are sought by naturalists, since they possess high aesthetic appeal, press easily, and can be memorized for a defined geographic area as there are far fewer species than flowering plants.

Ferns have been an important subject in the decorative arts from antiquity to the present. They have been used in motifs on ceramics (including fine porcelain), fabrics, glass, metalwork, paintings, and plastics. Pressed fern fronds become impressive crafts once they are mounted and framed, and they produce handsome herbarium specimens.

The human desire to explain phenomena through legends, folklore, myth, and symbolism extends to the ferns. The lack of flowers and seeds in ferns confused people and led to the development of superstitions concerning these plants.

Medieval people believed that ferns had seeds but they were invisible and that these "seeds" possessed magical powers of many kinds. Legends indicated that fern "seeds," if they could be found, were an essential ingredient of numerous desirable potions. Since these "seeds" were so important, myths and legends were developed that explained how and when they had to be collected to be most valuable and useful. Midsummer Eve was believed to be the proper time to gather fern "seeds," since this was the only period when ferns could bear "fruit." Gathering or picking ferns on this eve endowed the person with supernatural powers and good luck. The supposed magical powers associated with ferns led to the practice of hanging fern fronds over doors and windows of homes and barns.

According to legend, St. Patrick placed a curse on ferns and that is why they never produce flowers. Consumption of the sap found in fern stems was supposed to greatly extend one's youth. When ferns were included in medieval and later art they symbolized humility, sincerity, or frankness. Ferns can be seen in the background or border, or sometimes held in subjects' hands, in numerous pieces of art. Fern fronds were illustrated in English heraldry, coats of arms, and family crests.

Many fern species have specific legends or symbolic meanings associated with them. Such meanings in fables and legends have been incorporated into some of the scientific or common names applied to specific fern species (Ferguson 1959; Jobes 1962; Dunbar 1989).

The arrival of Europeans in what is presently New Jersey took place in the early seventeenth century. The English navigator Henry Hudson, sailing the *Half Moon* under the Dutch flag, recorded statements about the New Jersey landscape in the vessel's log in 1609. The earliest European settlements in the state were along the Delaware River by the Swedes between 1638 and 1640 and along the Hudson River by the Dutch in the 1640s. In 1650–1680 Swedish, Dutch, and English settlements were located along or near waterways throughout the state. England in 1664 claimed New Jersey and made it the fourth of the original 13 English colonies.

The influx of European settlers into New Jersey produced drastic changes in the natural landscape. The earliest census in 1726 recorded a population of 32,442; in 1784 it was 149,435; and there has been nonstop growth throughout the nineteenth and twentieth centuries. This population increase, along with extensive farming and industrial development, has greatly altered the appearance of the natural landscape and species distribution in New Jersey. Even with these changes, the pteridophytes are one segment of the floristic diversity that occupies most of the habitats found within the state. Thus, an understanding of the biology, habitat requirements, and geographic distribution of this group of plants provides insight concerning floristic changes that have occurred and may take place in New Jersey. The dynamics of these pteridophytes has aroused the interest of many amateur and professional botanists from near and far.

John Bartram, a botanist from Philadelphia, while collecting plants in 1736 in southeastern New Jersey, explored cedar swamps along the Great Egg Harbor River and found two species of ferns (*Osmunda*). Sometime during the 1740s he collected *Lycopodium* species growing in the sands of the "Jersey desert" (Pinelands) (Berkeley and Berkeley 1982).

Peter Kalm left Sweden and arrived in Philadelphia in September 1748. He explored the colonies for some 2½ years and spent half his time in New Jersey. Most of his New Jersey explorations were along the Delaware River in what are presently Burlington, Camden, Gloucester, and Salem counties. He collected specimens of *Osmunda cinnamomea* and *Osmunda regalis* in New Jersey in 1749 or 1750 (Juel and Harshberger 1929; Kalm 1972).

André Michaux sailed from France and arrived in New York City on November 13, 1785. He made his first American plant search 3 days later in northeastern New Jersey and sent numerous New Jersey plant specimens and seeds to Paris in December 1785. A special act of the General Assembly of New Jersey on March 3, 1786, permitted Michaux to purchase land for a botanical garden in the name of the king of France at Bergen Woods (in what is now Hudson County). This botanical garden was started by Michaux on March 28, 1786, to grow northern species for export to France. In 1796 he returned to France and began writing the first relatively complete American flora, *Flora Boreali Americana,* published in Paris in 1803. In the second of two volumes, 14 genera and 41 species of pteridophytes were listed. Many of these included New Jersey in the stated geographic range. Two species, *Nephrodium rufidulum* (= *Woodsia ilvensis*) and *Ophioglossum vulgatum,* are documented as New Jersey specimens collected by him and presently in the herbarium of the Musée Nationale d'Histoire Naturelle in Paris (Morton 1967; Ewan 1974).

Frederick T. Pursh (originally Pursch) left Dresden, Saxony, then part of Prussia, in 1791 and embarked for Baltimore. He became manager of "Woodlands" (1802–1805), a botanical garden owned by William Hamilton near Philadelphia. In 1806 Pursh was hired to assist Professor Benjamin S. Barton of the University of Pennsylvania in the collection of plants. In 1812 Pursh sailed to England, where he completed his *Flora Americae Septentrionalis,* published in London in 1814. He listed 27 genera (14 in New Jersey) and 93 species (42–48 in New Jersey) of pteridophytes. Four species had New Jersey sites specially cited, with the most famous taxon being *Schizaea pusilla.* Pursh, in the company of others, observed this species in a bog at Quaker Bridge, and he later described it as a new species in 1814 (Pursh 1814; Ewan 1952). Additional comments regarding the discovery are discussed under *Schizaea pusilla.*

William P. C. Barton, from Philadelphia, published *Compendium Florae Philadelphicae* in 1818, which included descriptions of indigenous and naturalized plants within 10 miles of Philadelphia. Listed in this book were 35 pteridophytes, of which 12 were cited as growing in New Jersey (Barton 1818).

Thomas Nuttall, who had no formal training in botany, arrived in Philadelphia from England in 1808 and worked as a printer. Soon after his arrival he associated with several naturalists, and within two months he was also employed by Professor B. S. Barton to collect plants. His New Jersey collecting included several different habitats. In 1818 he published the first flora written in English and published in the United States: *The Genera of North American Plants. . . .* He specifically referred to New Jersey locations for his genera *Dicksonia* (*Dennstaedtia*), *Isoetes, Lygodium, Schizaea,* and *Woodwardia* (Nuttall 1818).

John Torrey, M.D., was born in New York City in 1796. William Torrey, his father, was the financial officer of the New York State Prison, then located in New York City. He arranged for John to study botany from 1811 to 1815 with Professor Amos Eaton, an inmate who had been convicted of forgery and fraud. John later studied medicine and botany at the College of Physicians and Surgeons in New York City. He taught chemistry at

Princeton University for 24 years. He frequently visited the Pinelands and, in 1832, introduced Asa Gray, then 22 years old, to the plants of this region. Asa Gray moved into the Princeton home of the Torrey family in 1833 and became John's field assistant. In 1819 Torrey with two others published the *Catalog of Plants Growing Spontaneously Within Thirty Miles of the City of New York.* In this book the authors stated that Hoboken and Weehawken, NJ, as well as neighboring marshes, afforded excellent plant collecting. The cedar swamp near New Durham (Hudson County) became a significant botanical collecting site. A total of 44 pteridophyte taxa were listed for the flora, with 15 from New Jersey locations being specifically cited (Torrey et al. 1819).

Peter D. Knieskern, M.D., moved to Manchester (Lakehurst), NJ, from New York State and worked for William Torrey (John's brother) as postmaster of Manchester. In 1857 he published a catalog of plants of Monmouth and Ocean counties in which 19 pteridophytes were listed (Knieskern 1857).

Oliver R. Willis, Ph.D., lived in White Plains, NY, and was an instructor of natural science at the Alexander Institute. He published a catalog of plants growing in New Jersey. In the 1874 edition he listed 42 pteridophytes, and in the 1877 revised edition he listed 61 (Willis 1874, 1877).

Nathaniel L. Britton, Ph.D., upon being awarded an engineering degree from the School of Mines, Columbia College, in 1879, became an assistant in geology for Professor J. S. Newberry at Columbia. Britton's mentor recommended to Professor George H. Cook, Rutgers College, New Brunswick, and director of the Geological Survey of New Jersey that he hire Britton to be the botanist for the survey office. This became on-the-job botanical training for Britton, and the experience he gained with this survey led to his increasing interest and publications. Britton's manuscript for the preliminary plant catalog of New Jersey was published in 1881. Columbia College conferred the degree of Doctor of Philosophy on Britton in 1881 based largely upon the merit of this catalog (Howe 1934). Britton's completed New Jersey catalog (published in 1889) listed 58 species and 13 varieties of pteridophytes for the state.

During the last half of the nineteenth century botanical monographs became part of the scholarly publications written in the United States. The first two pteridophyte generic monographs written by American botanists were published in 1882 for *Isoetes* by G. Engelmann and in 1897 for *Ophioglossum* by E. G. Britton (see R. Tryon 1969). New Jersey specimens cited in these publications provided a significant portion of the research material studied by these two monographers.

An early report concerning *Isoetes* (*I. lacustris*) native to North America appeared in Pursh's *Flora Americae Septentrionalis* (1814). G. Engelmann described the species *Isoetes riparia* in Braun's (1846) account of North American *Isoetes.* He indicated that this new species grew in saline water between the high- and low-water lines, and he designated a herbarium sheet collected by William Zantzinger in 1844 from the shore of the Delaware River below Philadelphia as the type specimen. An earlier specimen of this taxon, originally designated as *I. lacustris,* was collected at approximately this site and reported by T. Nuttall in 1818. Engelmann's monograph (1882) included 12 species and 14 varieties, with several citations, including *I. riparia,* from New Jersey.

The second American generic pteridophyte monograph, published by E. G. Britton (1897) for *Ophioglossum* included *O. vulgatum,* collected and described by Michaux from New Jersey. The discovery in Cape May County of a new taxon, later named *O. arenarium* by E. G. Britton (1897), led to her revision of the genus; see our Comments in the discussion of *O. pusillum.*

Witmer Stone, while curator at the Academy of Natural Sciences in Philadelphia, published in 1911 an annotated, comprehensive, and illustrated floristic study of southern New Jersey, including the entire Pinelands. The extensive distribution and habitat records reported in this book provided basic phytogeographic data for 65% of the state. Many members of the Philadelphia Botanical Club had collected valuable herbarium specimens cited in his book. The thousands of excellent specimens collected by Bayard Long of Philadelphia, in addition to his diligent searching of undercollected southern New Jersey counties, provided an indispensable source of distribution and morphological variation information for Stone's book as well as later botanical studies that included New Jersey. Stone (1911) included 8 families, 20 genera, 50 species, and 4 varieties of pteridophytes. Another important source of botanical information on the region included in Stone's study was Louis E. Hand of Vincentown, NJ, a mentor of many naturalists studying the flora of southern New Jersey.

Norman Taylor, while a curator at the New York Botanical Garden and later at the Brooklyn Botanic Garden, studied the flora of the vicinity of New York City, which included the entire state of New Jersey. Frequently searching for plants with members of the Torrey Botanical Club, he concentrated on the northern counties of New Jersey. The pteridophyte treatment in his book (Taylor 1915) included 10 families, 28 genera, 82 species, and 13 hybrids of *Dryopteris.*

Minton A. Chrysler and James L. Edwards published the first fern book for the state in 1947. Dr. Chrysler was professor of botany at Rutgers University (1923–1946), and Mr. Edwards was a professional engineer and premier amateur pteridologist; they were an excellent combination. For many years Edwards diligently searched the entire state for pteridophytes and collected many quality herbarium specimens. He collected and studied such difficult genera as *Botrychium, Dryopteris, Equisetum, Lycopodium,* and especially *Isoetes.* His information and specimens have proven extremely valuable in the preparation of our book. Chrysler and Edwards (1947) listed 8 families, 29 genera, 78 species, 16 varieties or subspecies, 3 hybrids of *Asplenium,* and 13 hybrids of *Dryopteris.*

The decade of the 1970s included an increased awareness concerning rare plants in the United States. Several months before the U.S. Congress passed the Endangered Species Act of December 1973, Fairbrothers and Hough (1973) published an annotated list of rare and endangered New Jersey plant species. The interest in rare plants has led to some 20 years of conferences, symposia, field studies, and publications that have generated information about the present geographic and habitat status of rare pteridophytes incorporated into the present book (see Fairbrothers 1989; Montgomery 1989).

In an earlier publication (Montgomery and Fairbrothers 1985), we reported the known distribution by county of New Jersey's pteridophytes, recognizing 15 families, 29 genera, 80 species, 7 varieties, and 28 hybrids. In the present book we recognize 16 families, 30 genera, 83 species, 10 subspecific taxa, and 32 hybrids of pteridophytes as occurring in New Jersey.

We conclude this prefatory historical account by noting that information concerning pteridophytes occurring in New Jersey has been accumulated for more than two and a half centuries by many of the prominent botanists who have collected and investigated plants occurring in the northeastern United States.

Fern Structure and Classification

Description of ferns, like other kinds of plants and animals, requires use of specialized words and phrases. While we have attempted to avoid the use of technical terms where more familiar words will suffice, some specialized terms are essential for clarity. The purpose of this chapter is to explain and illustrate the terms used in this book.

Differences Between the True Ferns and the Fern-Allies

This book examines both true ferns and the so-called fern-allies. Both groups formerly were included in the classification that placed all spore-producing vascular plants in the Pteridophyta. Once considered to be related, the fern-allies are now understood not to be related to the true ferns, as each originated from distinct lines of early land plants and has had a long separate evolutionary history. In modern classifications the true ferns belong to the division Polypodiophyta; the fern-allies belong to Lycopodiophyta and Equisetophyta.

The main differences between the ferns and the fern-allies lie in the origin of the leaves and the position of the spore-bearing structures (sporangia). Fern leaves arose in evolution as branch systems that became flattened and webbed to form leaves (fronds) with complex branching veins (the vascular system). In the ferns, spores are borne on the backs or margins of the leaves. Leaves of the fern-allies arose in evolution as epidermal flaps of tissue (enations) from the stem; they have a single unbranched midvein. In the fern-allies the spores are borne in sporangia either at the bases of the leaves where the leaf joins the stem, in Lycopodiophyta, or on specialized branch systems (sporangiophores), in Equisetophyta.

The term "fern-allies" including Lycopodiophyta and Equisetophyta is thus a misnomer, since these plants are not allied in evolution to Polypodiophyta, the true ferns. Unfortunately, no satisfactory alternative has been proposed to replace the term fern-allies. For historical reasons and convenience the groups are usually treated together, as they are in this book. Terms for the structures of the fern-allies are somewhat different than for those of the true ferns and will be presented in a later section of this chapter.

Polypodiophyta: The True Ferns

The division Polypodiophyta includes some 10,000 species of true ferns, arranged in modern classifications into about 33 families and 425 genera (Crabbe et al. 1975). The ferns are worldwide in distribution (except Antarctica), although the majority of species occur in the tropics.

Fern Structure

THE RHIZOME

Ferns (Fig. 1) grow from stems called *rhizomes,* which are underground in all of the species native to New Jersey. The rhizomes produce roots and leaves. If the rhizomes are *erect* in the soil or very short, they produce shuttlecocks of leaves (Fig. 2a); if they are *short-creeping,* they produce rows of leaves (Fig. 2b); if they are *wide-creeping,* they produce scattered leaves (Fig. 2c). These patterns are very useful in making a field identification.

THE FROND

The leaves of ferns are called *fronds* (Fig. 1). Many of the terms needed for fern identification concern the frond, since this is the most visible part

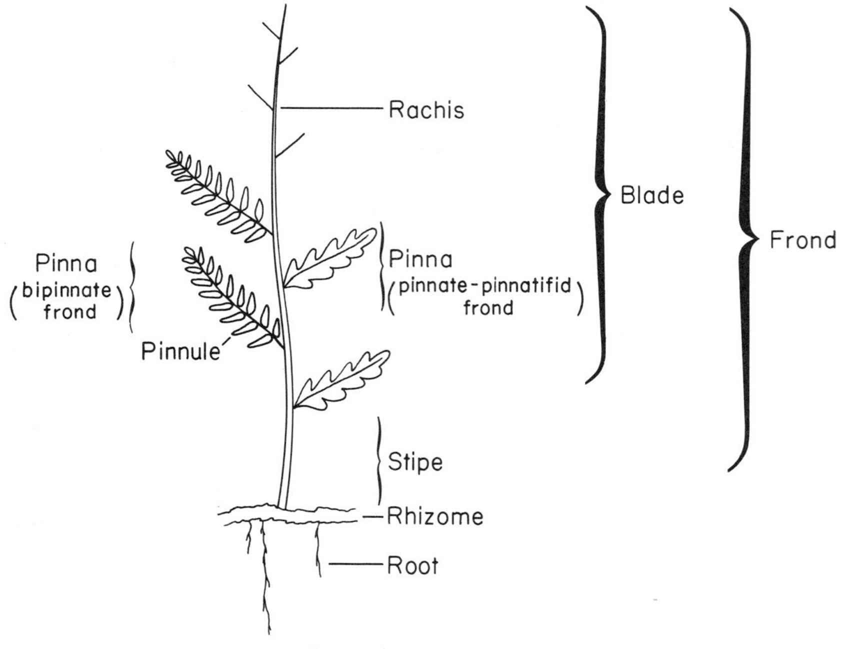

Figure 1. Parts of a fern.

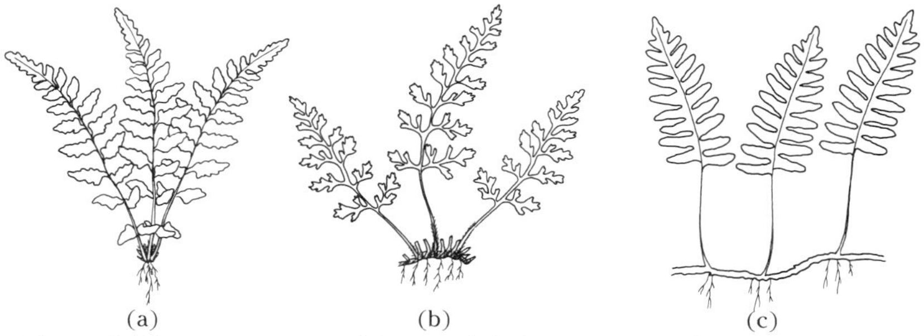

Figure 2. Rhizome patterns: (a) erect; (b) short-creeping; (c) wide-creeping.

of the plant. The flat green part of the frond is called the *blade* (Fig. 1). The stalk connecting the blade to the rhizome is the *stipe,* which is equivalent to the petiole in flowering plants. If the blade is divided (compound), the extension of the stipe into the blade is called the *rachis.*

The stipes contain *vascular bundles,* which transport water, minerals, and organic materials between the blade and the rhizome. The pattern of these bundles in the stipe is a characteristic useful for identification. To study the vascular bundles, we cut a cross section of the stipe near the base and examine with a 10 power lens or magnifying glass. Some of the patterns to be found are presented in Figure 3: single circular bundle, single U-shaped bundle, two strap-shaped bundles, and several (5–7) small round bundles arranged in a circle or U-shaped pattern.

For most of the ferns of New Jersey, the fronds emerge in the spring and die in the autumn; these are *deciduous* fronds. In some species the fronds persist through the winter and die back only as the new fronds emerge the following spring; these are *evergreen* fronds.

Frond Division

One of the most important sets of terms for identification concerns the cutting or division of the frond (Fig. 4). If the frond blades are not divided or lobed, they are *simple* (Fig. 4a). If the blades are deeply lobed but not cut all the way to the rachis, they are *pinnatifid* (Fig. 4b). If the blades are cut to the rachis forming separate parts, they are *pinnate* (Fig. 4c). The parts or *segments* are called *pinnae.* This distinction between pinnatifid

Figure 3. Vascular bundle patterns: (a) single circular bundle; (b) single U-shaped bundle; (c) two strap-shaped bundles; (d) several bundles.

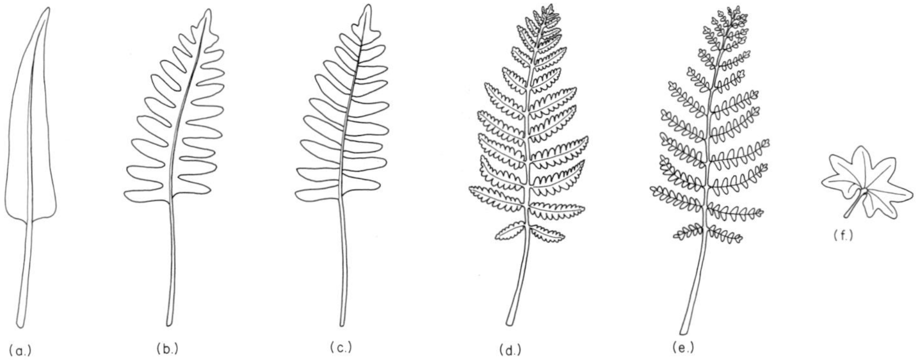

Figure 4. Frond division: (a) simple; (b) pinnatifid; (c) pinnate; (d) pinnate-pinnatifid; (e) bipinnate; (f) palmate.

(deeply lobed but tissue connecting the segments) and pinnate (cut into separate segments not connected by leaf tissue) is very important for identification. If a blade is pinnate and the pinnae are deeply lobed, the blade is called *pinnate-pinnatifid* (Fig. 4d). Note that this means the blade must be cut into separate segments and these segments in turn are deeply lobed. If a blade is pinnate and the pinnae are again cut into separate segments, the blade is *bipinnate* (Fig. 4e). The segments of a bipinnate blade are called *pinnules*. Note that only the smallest divisions of the divided blade are called segments: a *segment* is the smallest division of a blade. The blade can be divided more than twice, but the same terminology is extended. *Bipinnate-pinnatifid* means that the blade is bipinnate and the pinnules are deeply lobed. *Tripinnate* means that the blade is divided three times.

Most ferns that have pinnate blades have the pinnate arranged like the divisions of a feather (*pinnate*); in a few cases they radiate from a single point, and this is called *palmate* (Fig. 4f).

Blade Margin and Shape

The edges of the blades or segments are called the *margins*. The margins may be *entire* (Fig. 5a); finely saw-toothed, or *serrate* (Fig. 5b); or finely round-toothed, or *crenate* (Fig. 5c). There may be an enlarged lobe at the base of the pinnae; this is termed an *auricle,* and the pinnae are called *auriculate* (fig. 6). The rachis and its branches if the blades are pinnate are called the *axes* (sing., *axis*). The junction of the axis of a pinna and the main rachis is called the *axil* (Fig. 7). The pinnae may be *stalked* (attached by a short stalk) or *sessile* (with no stalk) (Fig. 8).

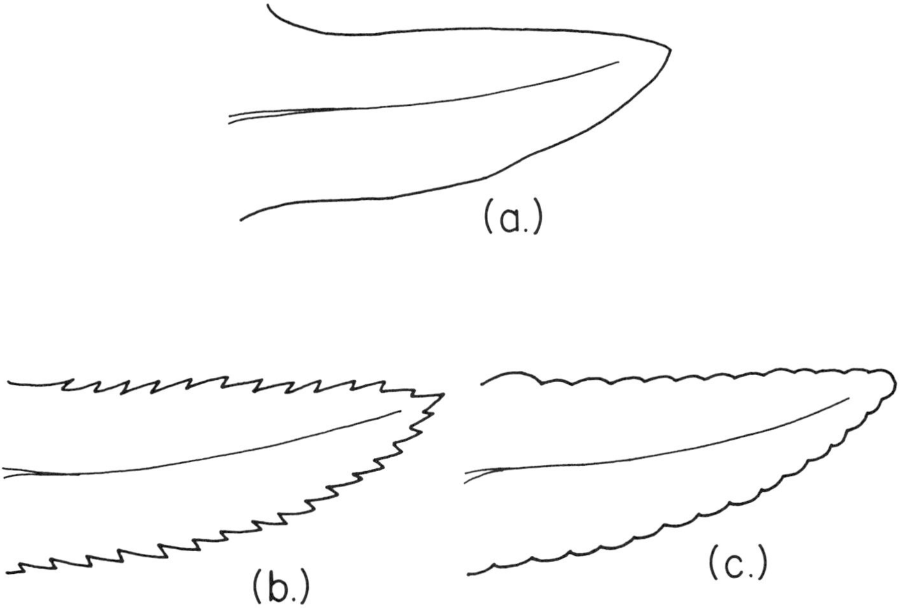

Figure 5. Margins: (a) entire; (b) serrate; (c) crenate.

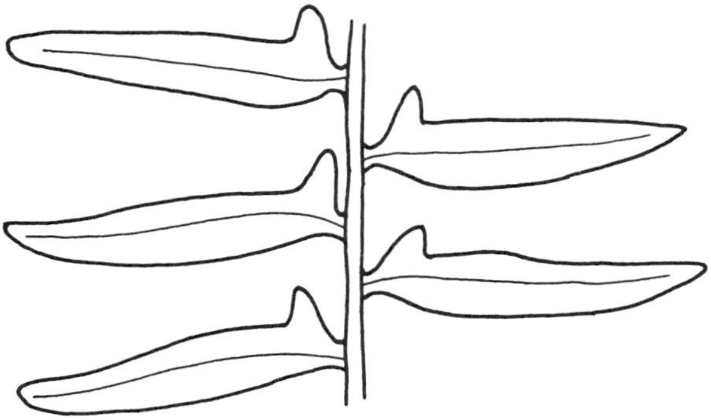

Figure 6. Auriculate.

The outline of the shape of the blades, pinnae, or pinnules may be important in identification. Terms used to describe the overall shape (outline) are *linear, lanceolate, ovate, elliptic,* or *triangular* (Fig. 9). Terms may be combined to indicate an intermediate condition, ovate-lanceolate, for example. The bases of the blades (or pinnae) may be *truncate, tapering,* or *cordate* (Fig. 10). The tips (*apices;* sing., *apex*) of the blades may be *acute, subacute,* or *obtuse* (Fig. 11).

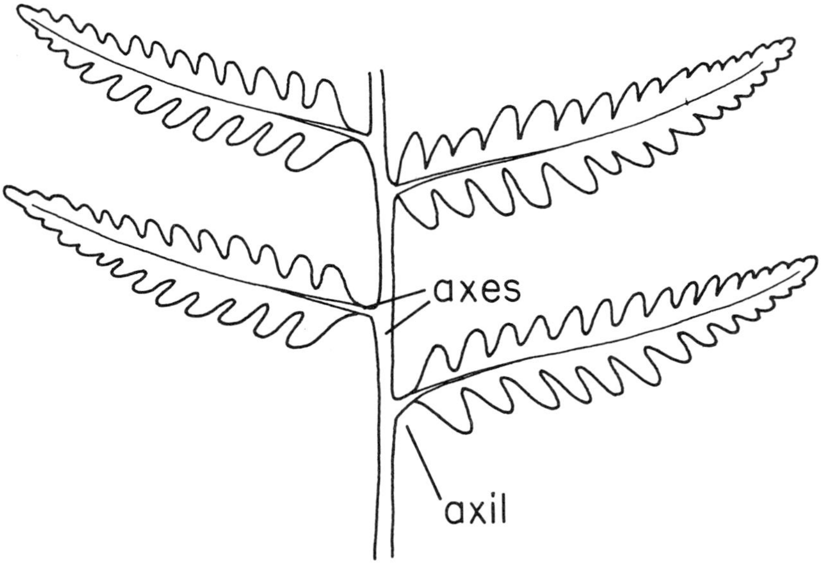

Figure 7. Axes and axil of a frond.

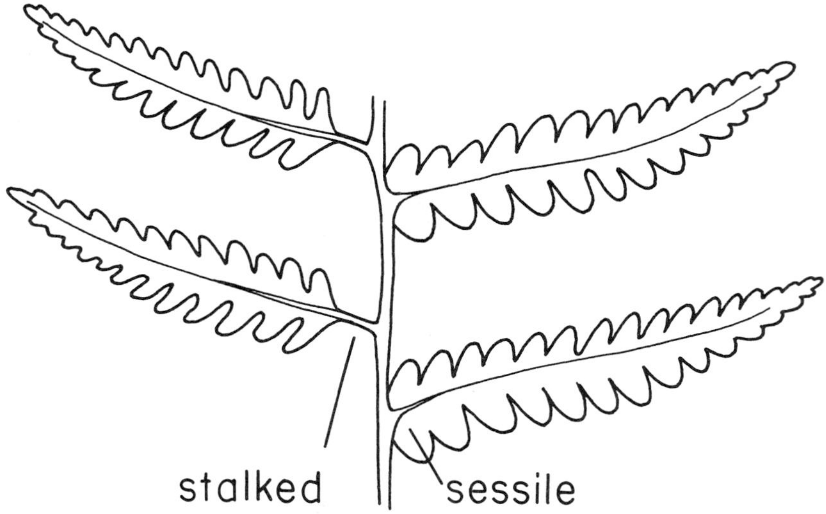

Figure 8. Stalked and sessile pinnae.

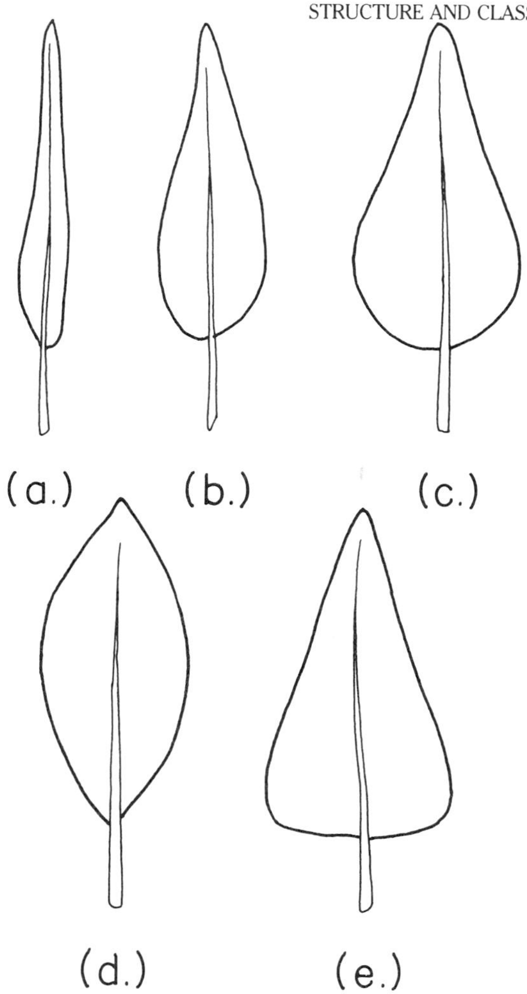

Figure 9. Blade shape: (a) linear; (b) lanceolate; (c) ovate; (d) elliptic; (e) triangular.

VENATION

In most ferns the veins in the segments are branched once or twice. If these branches do not rejoin to form a network, venation is *free*. If the veins join to form a network, venation is *areolate* (Fig. 12).

INDUMENT

Indument refers to the covering of the stipes or surface of the blades and includes hairs, scales, and glands. These characteristics usually must be

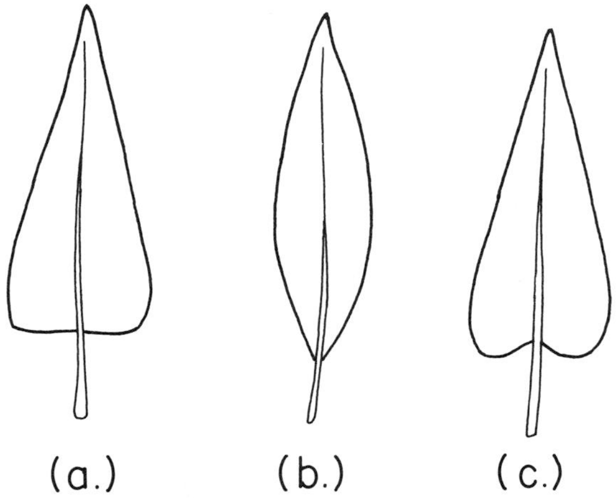

Figure 10. Blade base shape: (a) truncate; (b) tapering; (c) cordate.

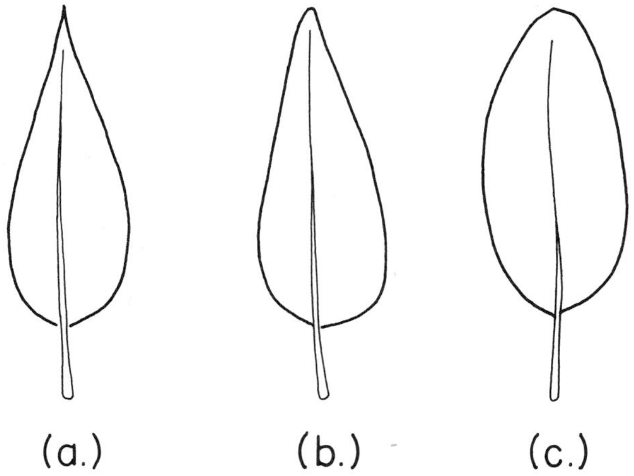

Figure 11. Blade apex shape; (a) acute; (b) subacute; (c) obtuse.

examined with a hand lens. *Hairs* are cylindrical structures that may be one-celled or have cross walls (jointed hairs). *Scales* are flat structures two or more cells across and several cells in length. Glands are modified hairs that secrete a fluid; in ferns they look like beads or hatpins on the blade surface. Surfaces that lack any indument are *glabrous.* Surfaces with hairs are designated hairy, or termed *woolly* if the hairs are very dense and matted (other terms are often used for different types of hairs,

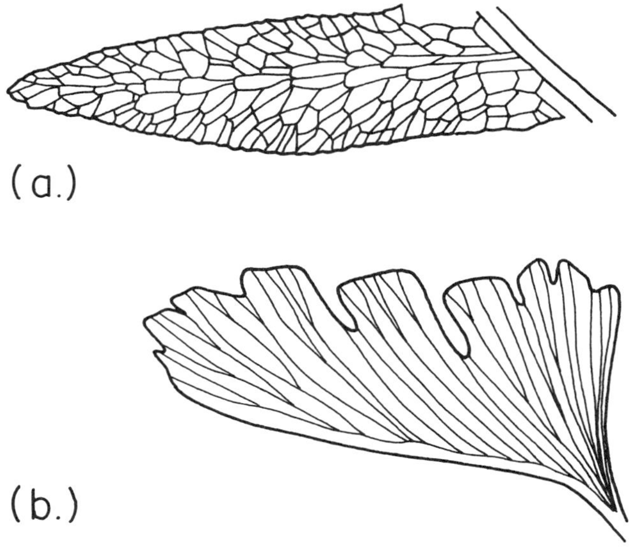

Figure 12. Venation patterns: (a) areolate; (b) free.

but will not be used in this book). Surfaces with scales are designated as scaly. Note that ferns may be hairy, or scaly or both.

Spores and Sori

Spores in ferns are borne in small cases called *sporangia.* In most but not all ferns the sporangia are grouped into clusters called *sori.* The sori are the brown-to-black spots that are seen on the backs of the mature fronds of ferns; the individual spores are too small to be seen without magnification. For purposes of fern identification (and classification), the important features are the location, shape, and structure of the sori. Structures including spores and details of the gametophyte are also important in classification but cannot be studied except at high magnification, so they are not discussed here.

Fertile and Vegetative Fronds

All of the fronds of a fern plant may be structurally alike and, at least potentially, bear spores. This condition is termed *monomorphic* (one form). Some fern species have fronds that bear the spores, the *fertile* fronds, different in size, shape, or structure from the leaves that do not bear spores, the *vegetative* fronds. Ferns that have different fertile and vegetative fronds are termed *dimorphic.*

SORUS SHAPE AND POSITION

The sori may be round, oval (elliptic), or elongate (linear) in shape. They may be located at the margins of the fertile fronds or segments, or they may be dorsal, on the backs of the segments. If dorsal, the sori are located near the midveins, near the margins, or medial between the midveins and the margins.

INDUSIUM

In many ferns the young sori are protected by small flaps of epidermal tissue called *indusia.* The indusia cover the young sori and upon drying expose the sori as they mature; remains of the indusia are usually visible with a hand lens. The way in which the indusia cover the sori is important in fern identification.

When the indusia cover the sori from one side like an awning, they are termed *lateral indusia* (Fig. 13a); when they cover the sori from a central stalk like a parasol, they are termed *peltate indusia* (Fig. 13b); when they cover the sori from a stalk attached at one side like a parasol with a V-shaped rip in it, they are termed *reniform indusia* (Fig. 13c); and when they surround the sori from below like a teacup, they are termed *inferior indusia* (Fig. 13d). In ferns with marginal sori, the margins of the segments may fold or roll over to cover the sori, and these are termed *false indusia* (Fig. 13f). Sori lacking indusia of any type are termed *naked* (Fig. 13e).

The Fern-Allies

Lycopodiophyta: Structure

The division Lycopodiophyta includes the living genera *Lycopodium, Selaginella,* and *Isoetes,* which are nearly worldwide in distribution, and also *Phylloglossum,* from Australia and New Zealand. The first three genera occur in New Jersey. Fossil representatives of this group were more diverse and grew to tree size; they were among the important coal-forming plants of the Paleozoic era.

The stems in this group are underground *rhizomes* (or climbing as *epiphytes* in the tropics), and may be wide-creeping, short-creeping, or modified into an erect, thick *corm,* as in a *Crocus,* in *Isoetes.* The small leaves, usually only 1–20 mm long, but up to a meter long in aquatic species of *Isoetes,* are either arranged spirally or in opposite pairs on the stems. The spores are borne in *sporangia* at the bases of the leaves. The leaves with sporangia (fertile leaves) may be similar to vegetative leaves

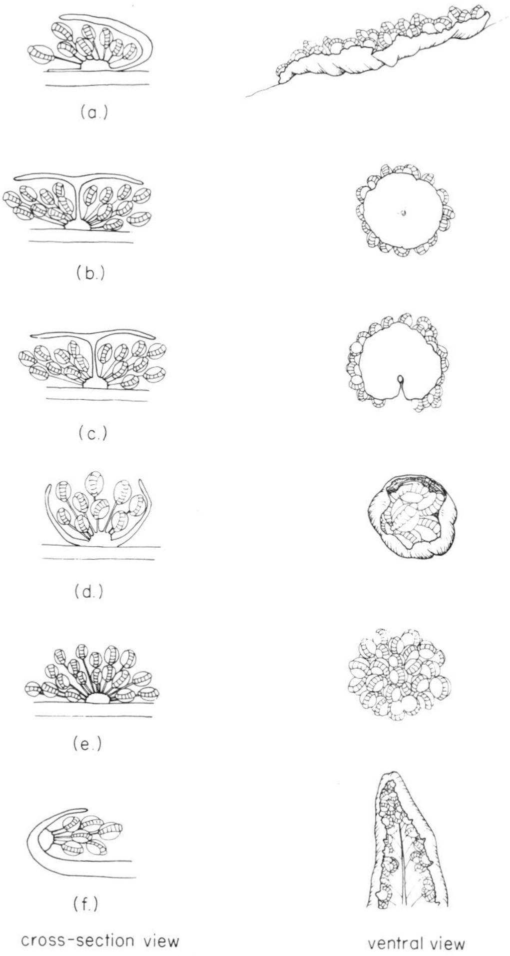

Figure 13. Indusium types: (a) lateral indusium; (b) peltate indusium; (c) reniform indusium; (d) inferior indusium; (e) naked (no indusium); (f) false indusium.

or they may be modified (usually shorter and thicker) and clustered at the tips of the branches into elongate *strobili* (sing., *strobilus*).

In *Selaginella* and *Isoetes* there are two kinds of spores: *microspores* (usually smaller), which contain the male gametophytes, and *megaspores* (usually larger), which contain the female gametophytes. The microspores are produced in *microsporangia;* the megaspores in *megasporangia.* The megaspores of *Isoetes* are very important in species identification and must be studied at magnifications of at least 40 power.

Equisetophyta: Structure

Modern plants in the division Equisetophyta all belong to the genus *Equisetum*. As in Lycopodiophyta, there were many fossil genera, some of which reached tree size. The structure of *Equisetum* is distinctive, and members of this genus are easily recognized.

The underground *rhizomes* are wide-creeping and produce erect stems, which may or may not branch. The erect stems are hollow, ribbed, and jointed. A circle of small leaves and sometimes branches is borne at the joints (*nodes*). The small, scale-like, single-veined leaves are fused into a *sheath* around the stem above the joint, with free *teeth* projecting from the sheath.

The spores are green and equipped with four strap-shaped *elaters*, which aid in dispersal. The spores are borne in sporangia arranged in a circle around a branch or *sporangiophore;* these are aggregated into *strobili*. All spores are alike in *Equisetum*.

Classification

As explained earlier, the traditional division Pteridophyta is now understood to comprise three separate evolutionary divisions: the lycopods, Lycopodiophyta; the horsetails, Equisetophyta; and the true ferns, Polypodiophyta. Within each of these divisions plants are divided into orders, orders into families, families into genera, and genera into species, according to perceived evolutionary relationships. Thus a family represents a group of genera related by evolutionary history.

The number of families of true ferns has undergone considerable revision in recent years. Traditionally the "more primative ferns" were placed in about ten families and the "advanced ferns" were placed in a single very large family, Polypodiaceae; the aquatic ferns, also considered "advanced," were placed in a separate family, Marsileaceae. For those interested in more details of the history of fern classification the review by Pichi Sermolli (1973) is recommended. The families of ferns were delimited primarily on the basis of sorus position and sporangium morphology.

In the last 40 years several classifications have been proposed for the true ferns. These utilize recent data from many disciplines including morphology, anatomy, cytology, genetics, chemistry, and ecology. The result has been the division of the Polypodiaceae into several families. The most recent complete classifications are those of Holttum (1949), Crabbe et al. (1975), Pichi Sermolli (1977), and Tryon and Tryon (1982). Although these classifications differ in the number of families recognized and in the disposition of certain genera, there is broad general agreement among them in the placement of most genera and the recognition of most family groups. The classification of Crabbe et al. (1975) is used in this book for the arrangement of the ferns and fern-allies of New Jersey because it is a fairly conservative, modern, and worldwide treatment and allows a degree of flexibility. The families and genera of New Jersey ferns and fern-allies are listed in Table 1.

Unfortunately many of the combinations of characteristics that define the modern families of ferns are not satisfactory for identification of

Table 1. Divisions, Families, and Genera of the Ferns and Fern-Allies of New Jersey, and the Number of Species and Hybrids Within Each Genus Recognized in New Jersey[a]

Division	*Family*	*Genus*	*Number of species/hybrids*
Lycopodiophyta	Lycopodiaceae	*Lycopodium*	11+/4+
	Selaginellaceae	*Selaginella*	2/0
	Isoetaceae	*Isoetes*	6/4
Equisetophyta	Equisetaceae	*Equisetum*	6/3
Polypodiophyta	Ophioglossaceae	*Botrychium*	7/0
		Ophioglossum	2/0
	Osmundaceae	*Osmunda*	3/0
	Schizaeaceae	*Schizaea*	1/0
		Lygodium	1/0
	Adiantaceae	*Cryptogramma*	1/0
		Cheilanthes	1/0
		Pellaea	2/0
		Adiantum	1/0
	Hymenophyllaceae	*Trichomanes*	1/0
	Polypodiaceae	*Polypodium*	1/0
	Dennstaedtiaceae	*Dennstaedtia*	1/0
		Pteridium	1/0
	Thelypteridaceae	*Thelypteris*	5/0
	Aspleniaceae	*Asplenium*	7/3
		Matteuccia	1/0
		Onoclea	1/0
		Athyrium	3/0
		Gymnocarpium	1/0
		Cystopteris	3/0
		Woodsia	2/0
		Polystichum	1/0
		Dryopteris	7/18
	Blechnaceae	*Woodwardia*	2/0
	Marsileaceae	*Marsilea*	1/0
	Azollaceae	*Azolla*	1/0
Total number	16	30	83/32

[a]Classification follows Crabbe et al. (1975).

field or herbarium specimens. Keys to families tend to be complex, with many families "keyed out" in more than one place; see, for example, the keys to families in Tryon and Tryon (1982) and Lellinger (1985). For this reason we have not provided a key to the families in this book.

Nomenclature

The assignment of scientific names to plants is called nomenclature. People apply common names to plants as a means of communicating with each other concerning the plants they observe and use. At various periods of time different names have been applied to the same plant or the same name to different plants. As books were written that included information about plants, confusion resulted from the lack of uniformity related to descriptions and names of plants. Even though some common names have had a very long existence, they are frequently ambiguous and, when transferred to different languages, have restricted usefulness. The oldest common names we now use were the common names used in ancient Greece and Rome. In medieval times and continuing well into the nineteenth century, plant names in botanical use were given a Latin name or a Latinized spelling. As the number of plants known to science began to increase greatly, a more precise naming system became essential. Before the mid-eighteenth century plant names were frequently composed of several words and were called polynomials. This complex descriptive name became cumbersome and the system had to be changed and standardized. In 1753 the binomial name replaced the polynomial name upon the publication of Linnaeus's *Species Plantarum.* Thus our present formal naming system dates from 1753, and nomenclatural rules have gradually become standardized worldwide through periodic legal revisions at international botanical congresses, presently held at 6-year intervals. These rules are published as the International Code of Botanical Nomenclature.

Each plant species is known by its generic name (a noun) and a specific epithet (usually an adjective or a noun acting as an adjective) and these two names form the binomial designated the species scientific name, always printed in italic. Each binomial must be followed by a third item, the author or authors who described and named the plant. The scientific name of the climbing fern is *Lygodium* (genus) *palmatum* (specific epithet). The full name is *Lygodium palmatum* (Bernh.) Sw. The abbreviation Bernh. stands for J. J. Bernhardi, who in 1801 named the plant *Gisopteris palmatum.* In 1806 O.P. Swartz transferred it to *Lygodium.* Thus these two men are the authors (not authority) of this binomial combination. The genus name *Lygodium* is derived from the Greek word meaning "flexible," a reference to the twining rachis. The specific epithet *palma-*

tum refers to spreading divisions of the fertile and sterile fronds that are comparable to the palm and fingers of the hand.

The generic name is a singular Latinized noun or word treated as a noun. It is always written with an initial capital letter. The specific epithet, written in lower case, may be derived from diverse sources, but it is often an adjective illustrating a diagnostic characteristic of the plant.

Some species have been further divided into subspecies (subsp.) or varieties (var.), and they bear a subspecific or varietal epithet, printed in italic. *Botrychium lanceolatum* (Gmel.) Angstr. subsp. *angustisegmentum* (Pease & Moore) Clausen demonstrates a subspecies designation. *Ophioglossum vulgatum* L. var. *pycnostichum* Fern. is an example of the correct designation for a variety.

Since the 1950s the study of hybrid complexes has greatly increased our knowledge as to the evolution and phylogenetics of pteridophytes. Sterile hybrids occur between pairs of species with no development of a hybrid complex; in other cases extensive reticulate hybrid complexes occur that create a challenge in terms of identification and nomenclature. Taxa of *Dryopteris* associated with the designation of hybrids will demonstrate the nomenclatural procedure as applied to hybrid taxa. *Dryopteris clintoniana* × *intermedia* is a formula that designates the two parental species involved in the hybrid. *Dryopteris* × *dowellii* (Farw.) Wherry indicates the use of a binomial for naming this same hybrid. The hybrid *Dryopteris carthusiana* × *intermedia* was given the specific epithet *triploidea* by Professor Wherry and is thus named *D.* × *triploidea* Wherry. (Note that the genus name may be abbreviated if the reference is clear, but the specific epithet is never abbreviated.)

Nomenclature for ferns resulting from recent experimental studies has undergone extensive name changes. Nomenclature is one of the tools of taxonomists, and it must be used properly even if frequent name changes produce frustration for some people.

Hybrids

Hybrids are common among the ferns and fern-allies, and they account for some of the problems that are encountered in field identification. This chapter explains how hybrids arise, how to identify them, and their importance in the diversity and evolution of the ferns and fern-allies.

Species of ferns (and other plants) are usually genetically isolated from each other; that is, there are barriers that restrict the flow of genetic material from individuals of one species to individuals of another species. Species that are distantly related (different families or genera) usually cannot interbreed and produce viable progeny. Within some genera, however, interbreeding between closely related species does occur, the result being hybrid plants. A hybrid is formed when one species crosses with another.

In New Jersey 32 hybrids have been reported in five genera, representing 25% of the pteridophyte taxa (species, hybrids, and subspecific taxa combined). Most hybrids are rare and will be encountered infrequently in the field. A few, especially in *Dryopteris* and *Lycopodium,* are fairly common; these are likely to lead to confusion. Some hybrids make excellent garden plants, since they are often larger than the parent species and in some cases combine desirable horticultural characteristics.

Fern hybrids are usually sterile; that is, they are unable to reproduce sexually, although many reproduce by vegetative propagation. This sterility results from the behavior of the chromosomes during the cell division that results in the formation of spores (meiosis). For this cell division to occur normally there must be corresponding, or homologous, pairs of chromosomes in the cell (one from each parent). The normal sexual pattern is for a sperm from an individual with chromosome complement A to fuse with an egg from another individual with complement A to form a zygote that is AA, which produces the sporophyte. In meiosis the chromosomes of the two A complements pair, separate, and form the spore cells, again each with complement A. (Note that these A's represent a whole set of chromosomes, not genes or individual chromosomes; there are few to very many chromosomes per set.)

If a sperm cell with complement A from one species fertilizes an egg with complement B from a different species, the result is a zygote that is AB. If the parental species are not closely related, this zygote usually dies or is not formed at all and so no hybrid results. If the parental species are similar enough genetically for the zygote to be viable and functional, a hybrid individual results. At the time of spore formation, the chromosomes from the A complement are unable to pair properly with those from the B complement. This results in spore abortion; the spores do not form and cannot germinate, and they often lack cell contents. These aborted spores are the most important clue in the detection of a hybrid fern plant. Chromosome counts can be used to identify hybrid plants, but they are difficult to accomplish for most people and impractical in the field. Spores can be studied with a dissecting microscope at approximately 40-power or, with some experience, with a 15-power hand lens. Normal spores of ferns and fern-allies are spherical, tetrahedral, or more or less bean-shaped (bilateral). No matter what the shape, normal spores from a species are all the same shape and size; under magnification they appear as tiny marbles or bean seeds. Aborted spores are very irregular in shape (shrunken and shriveled) and size. A few appear nearly normal, some look like basketballs, and they are mixed with many irregular-shaped particles. Under magnification (40-power or higher) aborted spores look like dirt or small soil particles.

Although aborted spores and chromosome counts are the definitive ways to recognize hybrids, there are other clues to aid recognition in the field. Hybrid plants look different (usually there are parent plants in the vicinity for comparison). The morphology of the hybrid plant is not typical for any species. It should be emphasized that becoming familiar with the species in the flora is essential before hybrids can be systematically hunted (Montgomery 1982a). Hybrids are often represented by one or two unusual individuals in a population. If such an unusual individual is found, it may be a hybrid and the spores should be checked. Hybrids are usually intermediate in morphology between the different characteristics of their parents. For example, if one of the parents has evergreen leaves and the other does not, the hybrid may be partially evergreen (leaves persisting into the early winter but eventually dying). This intermediate morphology makes the hybrid plant appear different from either of the parents and provides important clues in identification of the parents. Hybrids are best identified in this way: look for salient characteristics of the various potential parent species in the genus.

Hybridization has played a significant role in the evolution of ferns (Barrington et al. 1989). A hybrid with aborted spores cannot reproduce sexually because the chromosomes do not pair at meiosis; thus it is an evolutionary dead end. If the chromosomes of the hybrid are all doubled, however, the situation is changed. Thus, if A $\times$ B $\rightarrow$ AB (sterile hybrid) becomes doubled ($\rightarrow$ AABB), then such an individual with four sets of chromosomes, called a tetraploid, can behave as a fertile species because the AA and BB chromosome complements can pair normally. This individual can reproduce sexually, produce normal spores, and fertile offspring. Many fern species are believed to have arisen in this way (Wagner 1971; Barrington et al. 1989). Because tetraploid species can reproduce sexually, they can also hybridize with the parents that produced them (backcross). The tetraploid AABB can form gametes AB, and fertilize an A gamete from parent AA. This results in a triploid (three sets of chromosomes) hybrid AAB. Triploids are always sterile since there are never pairs for each complement of chromosomes. Beginning with the pioneer investigation of Manton (1950) much has been learned about fern evolution by studying the behavior of fern chromosomes. In the triploid backcross AAB above, for example, both pairs (AA) and single chromosomes (B) are formed at meiosis, indicating that the species with the A complement is one parent of the AABB species. For example, *Asplenium montanum* and *Asplenium platyneuron* are diploid species, each with a chromosome complement of 72 chromosomes (36 pairs). A hybrid has been formed between these species (it would have 72 single chromosomes) and has doubled its chromosome complement to form the fertile tetraploid *A. bradleyi* with 144 chromosomes, 72 from *A. montanum* and 72 from *A. platyneuron*. Further, *A. bradleyi* backcrosses with *A. montanum* to produce *A.* × *wherryi,* a sterile triploid with 36 single chromosomes from *A. platyneuron* via *A. bradleyi* and 36 pairs. The pairs originate from *A. montanum,* one complement from the *A. montanum* parent, the other via the *A. bradleyi* parent, that is, AAB.

Several fern genera in the eastern United States exhibit a pattern of evolution based on diploid species crossing to form sterile hybrids, followed by chromosome doubling to form tetraploid fertile species, and even crossing of the tetraploid with a distinct diploid species followed by chromosome doubling to form a hexaploid (6×) fertile taxon. Such a pattern of evolution is called reticulate evolution, following the early study of Wagner (1954) in *Asplenium.* Reticulate evolution is evident in *Asplenium, Dryopteris, Isoetes,* and *Lycopodium* among genera occurring in New Jersey. Further discussion of the results and significance of reticulate evolution is presented by Wagner (1971) and Lellinger (1985). Although most hybrids are relatively rare in New Jersey (some have been reported only once in the state), we have included all hybrids known from New Jersey in the generic treatments. The more common hybrids, which have five or more records for the state, are included in the keys, described, and illustrated. Rarer hybrids are briefly described and herbarium records listed, but no maps or illustrations are presented for these taxa. The number of hybrids known from New Jersey represents about one-fourth of the pteridophyte taxa from the state. This relatively high percentage is the result of New Jersey's geographic location and diverse topography, which gives the state a rich flora for its size, pteridophytes included (see the chapter on ecology and distribution, below). In genera such as *Dryopteris* (18 hybrids known from the state), *Lycopo-*

dium (at least 4 hybrids), *Isoetes* (4 hybrids), *Asplenium* (3 hybrids), and *Equisetum* (3 hybrids), the geographic ranges of species from north and south or mountains and coastal plain overlap in New Jersey, promoting the formation of hybrid taxa. This creates fertile ground for biological investigation (Wagner 1971; Montgomery 1976) and identification problems for the field naturalist.

Keys

A taxonomic key is a device whereby a choice is provided between two contrasting or contradictory statements (a couplet). The user chooses the one that matches the specimen to be identified; this leads to another choice or to the identification (name) of an organism. Various types of identification keys are available, but all included in this book are indented dichotomous keys. To use such a key, read both of the choices numbered 1, then choose the one that best fits the specimen you are trying to identify. Follow the indicated number directly below and read both of the choices. Proceed in this fashion until you reach the name of a plant (genus or species). This should be the correct name for your specimen; however, it is essential to compare the illustration and description for verification; if something seems incorrect, check back through the key to locate the mistake (indented keys make this fairly simple).

A couple of caveats: Read both choices carefully; the first may seem satisfactory until you read the second. Don't guess; measure if required, and make sure you understand the terms (see the chapter on fern structure).

We present two keys for identification of taxa. The first is for the genera of ferns and fern-allies in New Jersey. This key requires mature, fertile material and leads to the name of a genus, where a key to the species and hybrids in that genus can be found. The second key does not require fertile material but requires reasonably mature vegetative material and careful observations of morphology. This second key includes only the ferns (not the fern-allies) and leads directly to species identification. It is based on the key published by Montgomery (1978). The first key is the easier of the two to use if fertile material is available; if not, try the second key. When using either key be prepared to backtrack and recheck the choices you have selected.

Key to the Genera of Ferns and Fern-Allies of New Jersey

1. Leaves in a circle and partly fused, forming sheaths around the stems; stems jointed . . . . *Equisetum* (p. 105)

1. Leaves not fused in a circle around the stem; stems not jointed. **2**

- **2.** Plants aquatic, rooted in ponds or streams; with short, thick corms; bearing tufts of long quill-like leaves with sunken sporangia in the leaf bases . . . . *Isoetes* (p. 90)
- **2.** Plants either not rooted aquatic or, if so, lacking basal corms and long quill-like leaves **3**
  - **3.** Leaves small, simple, with midribs only; sporangia in axils of fertile leaves **4**

4. Plants less than 3 cm tall; spores of two types, borne in 4-sided strobili .... *Selaginella* (p. 84)

4. Plants more than 2 cm tall; spores all alike, borne in round strobili or at the bases of vegetative leaves .... *Lycopodium* (p. 53)

3. Leaves (fronds) usually larger, with branching veins, simple to compound; sporangia on the back of leaves or on modified leaves **5**

5. Plants aquatic, either minute and floating, or rooted in mud or shallow water and with 4-parted fronds **6**

6. Plants floating on water surface, minute................. *Azolla* (p. 276)

6. Plants rooted in substrate; with 4-parted fronds arranged like a four-leaf clover .... *Marsilea* (p. 274)

5. Plants not aquatic, although sometimes on wet soil; not floating; fronds simple to compound, but not 4-parted............................... **7**

7. Spores produced on modified fronds or segments of fronds, either not green or much reduced in size or different in shape from the vegetative fronds.. **8**

8. Sporangia large (1–2 mm), on modified frond segments; plants with a single frond per year **9**

9. Fronds lobed to compound; sporangia in branched clusters .... *Botrychium* (p. 127)

9. Fronds simple and entire; sporangia in a double row on an unbranched stalk.......................... *Ophioglossum* (p. 142)

8. Sporangia minute (<0.5 mm); plants normally with two to several fronds.. **10**

10. Plants grass-like, with curling green vegetative fronds and erect fertile fronds .... *Schizaea* (p. 152)

10. Plants not grass-like nor with curling green vegetative fronds... **11**

11. Fronds climbing on other vegetation, pinnate; with forked palmate-lobed pinnae .... *Lygodium* (p. 156)

11. Fronds not climbing, simple to pinnately compound **12**

12. Sporangia borne in clusters on separate pinnae or on separate fronds, not borne on nor covered by leaf tissue........................... *Osmunda* (p. 146)

12. Sporangia borne on modified fronds or segments and covered by or borne on leaf tissue................ **13**

13. Sporangia enrolled in highly modified fronds distinct from the vegetative fronds **14**

14. Vegetative fronds only pinnatifid .... *Onoclea* (p. 212)

14. Vegetative fronds pinnate-pinnatifid .... *Matteuccia* (p. 210)

13. Sporangia not enrolled by leaf tissue, borne on modified vegetative fronds or parts of fronds.................................. **15**

15. Sporangia borne on modified terminal portion of vegetative frond ..... *Polystichum* (p. 236)

15. Sporangia borne on modified fronds distinct from vegetative fronds **16**

16. Sori distinct, elongate along midvein of segments; vegetative fronds pinnatifid.......... .... *Woodwardia* (p. 270)
16. Sori confluent, borne on modified frond margin; vegetative fronds pinnate to bipinnate ....*Cryptogramma* (p. 160)

7. Fertile fronds not differentiated from vegetative **17**
17. Sporangia borne along margins of fronds **18**
18. Sori distinct **19**
19. Indusia formed from modified frond margins; sori oblong, elongate ....*Adiantum* (p. 158)
19. Indusia cup-like around sporangia; sori round ....*Dennstaedtia* (p. 172)
18. Sori confluent into a marginal band **20**
20. Fronds 50–150 cm tall, with stout stipe; rhizome wide-creeping, deep underground ....*Pteridium* (p. 174)
20. Fronds 10–40 cm tall, with slender, wiry stipe; rhizome short-creeping, surficial **21**
21. Frond segments hairy; segments 2–5 mm long ....*Cheilanthes* (p. 162)
21. Frond segments glabrous; segments 10–30 cm long ....*Pellaea* (p. 164)
17. Sporangia borne on underside of fronds, not along margins **22**
22. Indusia present **23**
23. Sori elongated.. **24**
24. Sori parallel to segment midrib .... *Woodwardia* (p. 270)
24. Sori on oblique lateral veins **25**
25. Stipes wiry, with a single ×-shaped vascular bundle; fronds less than 40 cm tall .... *Asplenium* (p. 190)
25. Stipes fleshy, with 2 ribbon-shaped vascular bundles; fronds usually more than 40 cm tall .... *Athyrium* (p. 214)
23. Sori round **26**
26. Indusia attached by the side of (lateral) or beneath the sori (inferior) **27**
27. Indusia hood-like, attached at side; stipes lacking scales except at very base ....*Cystopteris* (p. 226)
27. Indusia attached beneath the sori; stipes scaly and/or hairy.............................. *Woodsia* (p. 232)
26. Indusia attached at the center and covering the sori (superior) **28**
28. Indusia circular; fertile pinnae terminal and narrower than the vegetative pinnae below . *Polystichum* (p. 236)
28. Indusia reniform; all pinnae alike—the fertile not narrower or restricted to the terminal part of the blade **29**
29. Stipes scaly throughout; midrib and segment axes with scales and sometimes glands but not with pointed hairs; stipe bundles 5......... *Dryopteris* (p. 238)

29. Stipes scaly only at base; midrib and segment axes with pointed hairs but not scaly; stipe bundles 2 .... *Thelypteris* (p. 178)

22. Indusia absent (inconspicuous in *Woodsia*)............................ **30**

30. Fronds pinnatifid; sori large (3–5 mm in diameter). *Polypodium* (p. 169)

30. Fronds pinnate-pinnatifid to bipinnate; sori smaller (1–3 mm in diameter).. **31**

31. Blades triangular, widest at the base **32**

32. Blades glabrous; pinnate above the basal pair of pinnae .... *Gymnocarpium* (p. 224)

32. Blades with pointed hairs on the lower surface; pinnatifid above the basal pair of pinnae .... *Thelypteris* (p. 178)

31. Blades lanceolate-ovate, widest above the base (indusia present, but difficult to see in older material)............... *Woodsia* (p. 232)

Vegetative Key to Ferns of New Jersey

1. Blades simple to pinnatifid, not cut into separate pinnae **2**

2. Blades simple, margins entire **3**

3. Vegetative blades curled and grass-like .... *Schizaea pusilla* (p. 152)

3. Blades flat and leaf-like.. **4**

4. Blades elliptic, tapering to the stipes ....*Ophioglossum* spp. (See genus key p. 142)

4. Blades lanceolate, cordate at the bases .... *Asplenium rhizophyllum* (p. 192)

2. Blades pinnatifid **5**

5. Veins free.. **6**

6. Segments pinnatifid; primary segments connected by a wing along the rachis .... *Thelypteris hexagonoptera* (p. 180)

6. Segments serrate to entire **7**

7. Blades irregularly cut, upper lobes shallow; segments crenate; fronds tufted .... *Asplenium pinnatifidum* (p. 194)

7. Blades regularly and deeply pinnatifid; segments entire; fronds arising singly .... *Polypodium virginianum* (p. 170)

5. Veins areolate **8**

8. Lobes alternate, margins finely serrate; middle lobes of blade the longest.......................... *Woodwardia areolata* (p. 270)

8. Lobes opposite or nearly so, entire or wavy-margined; basal lobes of blade the longest .... *Onoclea sensibilis* (p. 212)

1. Blades at least once-pinnate.. **9**

9. Blades with entire to serrate pinnae **10**

10. Pinnae with auricles **11**

11. Stipes scaly, with several bundles in cross section; fronds thick-textured .... *Polystichum acrostichoides* (p. 236)

11. Stipes smooth and shining above base, with one bundle in cross section; fronds thin-textured... *Asplenium platyneuron* (p. 196)

10. Pinnae lacking auricles.. **12**

12. Blades and especially stipes fleshy in texture; fertile segments very different from the vegetative and nearly always present. . . . . *Botrychium* spp. (see genus key, p. 127)

12. Blades and stipes not particularly thick or fleshy; fertile parts of frond (if present) much like the vegetative **13**

13. Pinnae elongate, entire **14**

14. Pinnae blunt; stipes dark brown or purple. **15**

15. Stipes shiny brown, nearly glabrous . . . . *Pellaea glabella* (p. 164)

15. Stipes rough-hairy, dull brown or purple. . . . . *Pellaea atropurpurea* (p. 166)

14. Pinnae acute; stipes green . . . *Athyrium pycnocarpon* (p. 216)

13. Pinnae nearly round, serrate. . . . *Asplenium trichomanes* (p. 198)

9. Blades with pinnae pinnatifid to pinnate (or more divided) **16**

16. Stipes forking at summit into equal branches; these pinnate. . . . . *Adiantum pedatum* (p. 158)

16. Blades pinnately branched throughout **17**

17. Blades elongate, climbing; pinnae forked into 2 palmately lobed pinnules . . . . *Lygodium palmatum* (p. 156)

17. Blades neither elongate nor climbing; pinnae not as above **18**

18. Blades broad-triangular, more than half as broad as long, the lowest pinnae largest and longest **19**

19. Fronds 0.5 m tall or more, coarse . . . . *Pteridium aquilinum* (p. 174)

19. Fronds less than 0.5 m tall, more delicate **20**

20. Middle pinnae connected by an angular wing along the rachis **21**

21. Lowest pinna pair not united by a wing; lower blade surfaces hairy and scaly . . . . *Thelypteris phegopteris* (p. 182)

21. All pinnae united by wings; lower blade surfaces sparsely hairy, not scaly . . . . *Thelypteris hexagonoptera* (p. 180)

20. Pinnae free except at tip of blade **21**

22. Stipes sparsely scaly, wiry . . . . *Gymnocarpium dryopteris* (p. 224)

22. Stipes glabrous, fleshy **23**

23. Blades herbaceous in texture, deciduous . . . . *Botrychium virginianum* (p. 128)

23. Blades leathery in texture, evergreen. . . . . *Botrychium dissectum* (p. 136)

18. Blades lanceolate to elliptic, less than half as wide as long. . . . **24**

24. Pinnae reduced to small wings at base of blades (blades tapering to both ends) **25**

25. Stipes dark brown, thick; fronds clustered; blades glabrous . . . . *Matteuccia struthiopteris* (p. 210)

25. Stipes green, thin; fronds scattered; blades hairy . . . . *Thelypteris noveboracensis* (p. 184)

24. Pinnae little reduced at base of blades **26**
26. Veins areolate .... *Woodwardia virginica* (p. 272)
26. Veins all free **27**
27. Fronds 15 cm long or less (mostly small rock ferns with tufted fronds) **28**
28. Stipes brown or purple-black throughout **29**
29. Blades bipinnate, segments not lobed.................. **30**
30. Stipes shining brown, nearly glabrous .... *Pellaea glabella* (p. 164)
30. Stipes dull brown, rough scaly .... *Pellaea atropurpurea* (p. 166)
29. Blades bipinnate-pinnatifid or tripinnate **31**
31. Stipes glabrous, or nearly so . *Asplenium bradleyi* (p. 202)
31. Stipes hairy or scaly **32**
32. Stipes jointed, leaving stubble of uniform old bases .... *Woodsia ilvensis* (p. 232)
32. Stipes not jointed, stubble uneven **33**
33. Stipe cross section with 2 bundles; blades and axes scaly but not hairy.................... .... *Woodsia obtusa* (p. 234)
33. Stipe cross section with one bundle; blades and axes with white to brownish hairs...... .... *Cheilanthes lanosa* (p. 162)
28. Stipes brown only at base **34**
34. Rhizomes creeping, fronds spaced apart (fertile fronds very different from the vegetative) .... *Cryptogramma stelleri* (p. 160)
34. Rhizomes erect, fronds tufted (fertile and vegetative fronds alike).. **35**
35. Segments alternate, few, wedge-shaped............... .... *Asplenium ruta-muraria* (p. 200)
35. Segments opposite, numerous, ovate .... *Asplenium montanum* (p. 204)
27. Fronds more than 15 cm long **36**
36. Fronds bipinnate, the segments not pinnatifid **37**
37. Fronds small, not over 15–20 cm; blade bipinnate only at base; segments thick-textured......................... .... *Pellaea atropurpurea* (p. 166)
37. Fronds larger, usually 50 cm or more; blades bipinnate throughout; segments thin textured .... *Osmunda regalis* (p. 150)
36. Fronds bipinnate-pinnatifid or more divided................. **38**
38. Fronds pinnate-pinnatifid, the segments or lobes of the pinnae entire **39**
39. Stipe bundle 1, horseshoe-shaped with incurved edges **40**
40. Tuft of hairs in axils of mature pinnae; pinnae subacute at tip ..... *Osmunda cinnamomea* (p. 146)

40. No tuft of hairs in axils of mature pinnae; pinnae rounded at tip. . . . . *Osmunda claytoniana* (p. 148)

39. Stipe bundles 2 at base of stipe. **41**

41. Veins of pinnules forking; pinnae with basal pinnules not reduced. . . . . *Thelypteris palustris* (p. 188)

41. Veins of pinnules not forking; pinnae with basal pinnules reduced **42**

42. Rhizomes wide-creeping, fronds spaced; rachis sparsely hairy . . . . *Thelypteris simulata* (p. 186)

42. Rhizomes short-creeping, fronds tufted; rachis densely hairy . . . . *Athyrium thelypterioides* (p. 218)

38. Fronds pinnate-pinnatifid or more divided, the lobes of the pinnae toothed, lobed, or pinnate **43**

43. Blades with the lower surface densely hairy; stipe bundle 1, circular . . . . *Cheilanthes lanosa* (p. 162)

43. Blades with the lower surface sparsely hairy, scaly, or glabrous; stipe bundles 2–5, or 1 and horseshoe-shaped. **44**

44. Rhizome wide-creeping; rachises finely hairy; base of stipes with 1 horseshoe-shaped bundle . . . . . *Dennstaedtia punctilobula* (p. 172)

44. Rhizome short-creeping; rachises glabrous or hairy; base of stipes with 2–5 bundles **45**

45. Cross section at base of stipe with 2 bundles. **46**

46. Stipes and rachises with no hairs or scales above base . . **47**

47. Fronds lanceolate, elongated, often with bulblets beneath upper part of blades; stipes pinkish . . . . *Cystopteris bulbifera* (p. 226)

47. Fronds ovate-oblong, not bearing bulblets; stipes green to straw-colored or brown **48**

48. Rhizomes extending beyond tuft of fronds; stipes pale, greenish; usually growing on soil in woods . . . . *Cystopteris protrusa* (p. 230)

48. Fronds clustered at ends of rhizomes; stipes straw-colored to brown; usually growing among rocks . . . . *Cystopteris tenuis* (p. 228)

46. Stipes and rachises with scales or hairs or both **49**

49. Cross section of stipe with 2 strap-shaped bundles; fronds usually stout, blades 10 cm or more broad **50**

50. Blades pinnate-pinnatifid, with narrow hair-like scales. . . . . *Athyrium thelypterioides* (p. 218)

50. Blades bi- to tripinnate, scaleless or with a few broad scales . . . . . *Athyrium filix-femina* (p. 220)

49. Cross section of stipe with 2 circular bundles; fronds slender, blades 8 cm or less broad . . . . *Woodsia obtusa* (p. 234)

45. Cross section at base of stipe with 5 small bundles **51**

51. Blades bipinnatifid to tripinnate at base, the segments with sharp-pointed prominent teeth. **52**

52. Rachises and minor axes glandular; fronds evergreen .... *Dryopteris intermedia* (p. 250)

52. Rachises and minor axes without glands; fronds deciduous .... *Dryopteris carthusiana* (p. 252)

51. Blades pinnate-pinnatifid to bipinnate only at base; the segments with round to somewhat acute teeth, or entire **53**

53. Blades leathery in texture, evergreen; scales at base of stipe in a dense tuft, orange-tan......................... *Dryopteris marginalis* (p. 240)

53. Blades herbaceous in texture, eventually deciduous; scales at base of stipe not tufted, light to dark brown............................... **54**

54. Lowest pinna pair triangular, broadest at base; scales light brown **55**

55. Pinnae turned at right angles to plane of frond (like venetian blind slats); fronds narrow; usually a few small evergreen fronds at base of plant .... *Dryopteris cristata* (p. 248)

55. Pinnae not twisted out of plane of blade; fronds not as narrow, but more variable; small evergreen fronds present or absent .... *Dryopteris clintoniana* (p. 246)

54. Lowest pinna pair ovate, narrowed at base; scales dark brown ... **56**

56. Fronds abruptly narrowed at tip; scales at base of stipe dark brown and shiny................ *Dryopteris goldiana* (p. 242)

56. Fronds tapering to tip; scales of stipe brown, often with a darker brown stripe, usually dull .... *Dryopteris celsa* (p. 244)

Ecology and Distribution

This chapter presents a general overview of the ecology and distribution of the ferns and fern-allies of New Jersey. The 114 species and hybrids of pteridophytes found within New Jersey are not uniformly distributed in the state: some occur in a single region, some are near the limits of their geographic range, some are restricted to specific types of soil or bedrock, and a few are found almost everywhere that vegetation occurs. This chapter reviews the physiographic provinces, surface geology, soils, climate, and vegetation, and how these factors affect the distribution of the ferns and fern-allies of the state.

New Jersey is a small state, forty-sixth in size among the 50 states, with an area of 7836 square miles. The population of more than $7\frac{1}{2}$ million, ninth among the 50 states, makes New Jersey the most densely populated state, with 1005 people per square mile. It surprises many people, including many New Jerseyans, to learn there are wild areas in the state. These areas are mostly located in the Pinelands or in the northwest corner of the state, away from the densely populated corridor between Camden and Ft. Lee.

New Jersey lies in the Mid-Atlantic region, bordered by the Atlantic Ocean and the Hudson River (and New York State) on the east, Delaware Bay on the south, the Delaware River (and the states of Delaware and Pennsylvania) on the west, and New York State on the north. The location of New Jersey is important in discussing the ferns and fern-allies of the state, because it is in New Jersey that the Appalachian Mountains are closest to the Atlantic Ocean, and the broad Coastal Plain, extending from the Gulf of Mexico northward, narrows and all but disappears. This proximity of the ridges of the Appalachians and the Coastal Plain is a contributing factor for the diversity of ferns and fern-allies in the state.

Climate is another important factor. Parts of New Jersey have a climate modified by the proximity of the ocean, which maintains higher temperature along the coast in winter and also brings coastal storms to much of the state. The ridges of the northwest have a more continental climate, with warm summers and cold winters; also snow is common and can be deep. Annual rainfall averages range from 40 inches in the southeast to 48 inches in the northwest, evenly distributed over the months. Annual snowfall totals range from 14 inches at Cape May to 50 inches in Sussex County. The lowest temperature recorded was −34°F in Riverdale, the highest was 110°F in Runyon (Tedrow 1986). The growing season ranges from 225 days at Cape May to 120 days in Sussex, and the frost-free period ranges from 200 to 140 days. Minimum temperature is more important for many species than growing season length, and the location and topographic features of New Jersey create a potential for a wide range in such temperatures (Parsons and Shulman 1978). The lowest minimum temperatures (−25° to −29°F) occur in the extreme northwest corner, with annual minima becoming progressively higher to the south and east. The highest minimum temperatures are recorded at Atlantic City and Cape May. Urban areas also have higher minimum temperatures than surrounding area. A zone of lower minimum temperatures occurs in the Pinelands, where radiational cooling is associated with flat topography, low urbanization, and increased distance from the ocean (Parsons and Shulman 1978).

New Jersey includes four physiographic regions: the Ridge and Valley, the Highlands, the Piedmont, and the Coastal Plain. The Coastal Plain can be divided into the Inner and Outer Coastal Plain (Wolfe 1977). These regions differ in geology, soils, vegetation, and the species of ferns and fern-allies that are found in each.

New Jersey was included by Braun (1950) in the Oak-Chestnut Forest Region and the Oak-Pine Forest Region. Even when proposed (1950) the name oak-chestnut was an anachronism, for the American chestnut had been gone from New Jersey forests for 30 years. This region includes all of northern New Jersey including the Ridge and Valley, Highlands, Piedmont, and Inner Coastal Plain, and was referred to as the Glaciated Section of the Oak-Chestnut Forest Region (even the part of the region never glaciated). The Outer Coastal Plain was mapped as the Oak-Pine Forest Region.

Robichaud and Buell (1973) identified 12 terrestrial habitat and plant community types in New Jersey, including six wetland types and six upland types. Their classification is more detailed and useful, and we have framed our discussion of the distribution of the pteridophytes with reference to the types of communities discussed by these authors.

The Ridge and Valley

The Ridge and Valley Province crosses northwestern New Jersey in Warren and Sussex counties (Map 1). This region consists of a series of parallel ridges and valleys trending northeast to southwest (Wolfe 1977). The Ridge and Valley Province is underlain by a series of early Paleozoic sedimentary rocks including conglomerate, sandstone, shale, and limestone in beds of various thickness that were tilted at the end of the Paleozoic Era. Subsequent differential erosion of hard (conglomerate and sandstone) and soft or soluble (shale and limestone) strata has produced the ridges and valleys, respectively, of this region. The ridges are underlain

Map 1. Physiographic provinces and limit of Wisconsin glaciation in New Jersey (from Wolfe 1977; Tedrow 1986).

by sandstones and conglomerates, which are acidic rocks. The valleys are underlain by limestone, which is calcareous (Map 2), and shale, which is usually acidic. The land is mantled with glacial drift of varying thickness. Glacial lakes are abundant; more than 65 of them remain in northern New Jersey. Almost all of the swamps of northern New Jersey are also products of disrupted drainage or filled glacial lakes.

Soils on the ridges are rocky, well drained, and subject to intense leaching (Tedrow 1986). Soils of the valleys are derived from glacial till, and are well drained to poorly drained depending on topography. Well-drained soils have a pH of 6.9–7.5; poorly drained soils, often developed on glacial lake beds (Great Meadows, for example) are more variable in pH.

Vegetation in the Ridge and Valley Province is influenced by slope and exposure and, therefore, is closely related to geology and soils. Vegetation on ridge tops, with hard sandstone or conglomerate rock and thin acid soils, is dominated by either chestnut oak forest or pitch pine–scrub oak forest (Robichaud and Buell 1973). Shrubs in these forests are primarily heaths; herbs are uncommon because of the rapid drying of the thin soil. There are few ferns on the dry acid ridge tops, although northeastern bracken (*Pteridium aquilinum* var. *latiusculum*) occurs in the oak-pine forests and hay-scented fern (*Dennstaedtia punctilobula*) grows along rock walls and forest edges.

Areas of exposed rocks occur on the ridge tops as a result of either extreme slope where soil does not accumulate (cliffs and ledges) or scouring by the glacier. Lichens, mosses, hairgrass (*Deschampsia flexuosa*), poverty grass (*Danthonia spicata*), sedges (*Carex* spp.), and early saxifrage (*Saxifraga virginiensis*) are typical plants of cliffs and rocky fields. Acid rock outcrops have some of New Jersey's rarer ferns, including Bradley's spleenwort (*Asplenium bradleyi*), lobed spleenwort (*A. pinnatifidum*), mountain spleenwort (*A. montanum*), and Trudell's spleenwort (*A.* × *trudellii*).

On drier south-facing slopes a mixed oak forest occurs, with many of the same species as in the ridge top forest; dogwood (*Cornus florida*) is important in the understory, with a few shrubs, grasses, sedges, and annual herbs. South-facing slopes tend to exclude pteridophytes because of the exposure and drying of the soil.

North-facing slopes are more moist, with a forest of sugar maple (*Acer saccharum*), tulip tree (*Liriodendron tulipifera*), white ash (*Fraxinus americana*), and basswood (*Tilia americana*), mixed with oaks. Fewer understory trees, more shrubs, and more perennial herbs are present (Robichaud and Buell 1973). Ravines, narrow valleys, and steep north-facing slopes have a forest dominated by eastern hemlock (*Tsuga cana-*

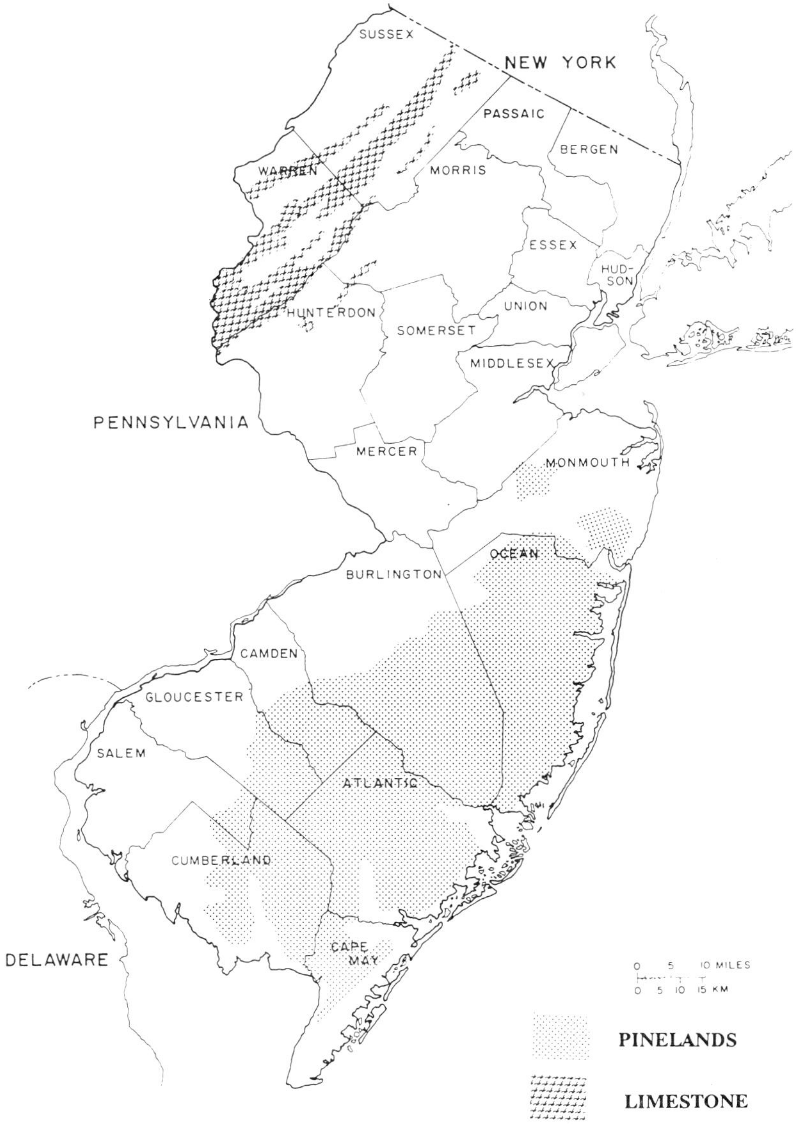

Map 2. Areas of limestone bedrock (Tedrow 1986) and approximate area of the Pinelands (Harshberger 1916) in New Jersey.

densis). Shrubs and herbs are not abundant because of the dense shade and acid soil caused by hemlock. North-facing slopes with more mesic forest conditions usually contain several ferns including New York fern (*Thelypteris noveboracensis*), northern lady fern (*Athyrium filix-femina*), marginal wood fern (*Dryopteris marginalis*), evergreen wood fern (*D. intermedia*), and Christmas fern (*Polystichum acrostichoides*). Evergreen wood fern and shining clubmoss (*Lycopodium lucidulum*) are also typical of hemlock ravines, along with common polypody (*Polypodium virginianum*) on exposed boulders. These cool, moist ravines may also contain northern species, including northern beech fern (*Thelypteris phegopteris*), and oak fern (*Gymnocarpium dryopteris*).

In the limestone valleys (Map 2) with calcareous soils, much of the natural vegetation has been removed for farming on the fertile soils. Remnants of vegetation indicate that this area was occupied by sugar maple and associated hardwoods characterized by many species of trees, a dense shrub layer, and abundant ferns and herbs (Robichaud and Buell 1973). Typical fern species are New York fern, northern lady fern, Christmas fern, marginal wood fern, maidenhair fern (*Adiantum pedatum*), southern beech fern (*Thelypteris hexagonoptera*), and interrupted fern (*Osmunda claytoniana*). Rarer species occasionally found in this community include Goldie's wood fern (*Dryopteris goldiana*), silvery spleenwort (*Athyrium thelypterioides*), glade fern (*A. pycnocarpon*), and ostrich fern (*Matteuccia struthiopteris*). Limestone ledges and boulders are occupied by ferns that are the best adapted to calcareous conditions. Walking fern (*Asplenium rhizophyllum*), bulblet fern (*Cystopteris bulbifera*), purple cliff-brake (*Pellaea atropurpurea*), and less frequently smooth cliff-brake (*P. glabella*), wall-rue (*Asplenium ruta-muraria*), and Scott's spleenwort (*A.* × *ebenoides*) are ferns restricted to calcareous soil in New Jersey.

Swamps may be dominated by shrubs including buttonbush (*Cephalanthus occidentalis*), alder (*Alnus serrulata*), and willows (*Salix* sp.) or by trees of red maple (*Acer rubrum*). A variety of shrubs and herbs occur, including sedges, skunk cabbage (*Symplocarpus foetidus*), and ferns. Northern bogs, which are always glacial features, may include open water and a mat of herbaceous vegetation, surrounded by characteristic shrubs and trees. Swamp ferns of the Ridge and Valley Province include royal fern (*Osmunda regalis*), cinnamon fern (*O. cinnamomea*), sensitive fern (*Onoclea sensibilis*), marsh fern (*Thelypteris palustris*), crested wood fern (*Dryopteris cristata*), Clinton's wood fern (*D. clintoniana*), and spinulose wood fern (*D. carthusiana*). Red maple swamps are the best habitats to find wood fern hybrids. Bogs, with their very acid soil, have few ferns, although marsh fern and cinnamon fern may occur around the edges, and Massachusetts fern (*Thelypteris simulata*), a species more typical of the

Table 2. Species of Pteridophytes Restricted to the Ridge and Valley (R) or Highlands (H) of New Jersey

Lycopodium annotinum (H)	*Thelypteris phegopteris* (R,H)
Lycopodium inundatum (R,H)	*Asplenium bradleyi* (R,H)
Isoetes lacustris (H)	*Asplenium pinnatifidum* (R,H)
Isoetes tuckermannii (R)	*Asplenium ruta-muraria* (R)
Equisetum variegatum (R,H)	*Athyrium pycnocarpon* (R)
Pellaea glabella (R,H)	*Gymnocarpium dryopteris* (R,H)
Trichomanes sp. (R)	*Cystopteris bulbifera* (R)

Coastal Plain, also occurs in bogs in northern New Jersey. Marshes and ponds are abundant, although many such habitats have been drained for agriculture, housing, or commercial development in northern New Jersey. Many of the small ponds and lakes in northern New Jersey contain one or two species of quillworts, including spiny-spored (*Isoetes echinospora*), Engelmann's (*I. engelmannii*), and lake quillwort (*I. lacustris*).

Abandoned fields and pastures are important plant communities in the Ridge and Valley Province. When agricultural land is abandoned it reverts to forest through a series of stages called succession. Abandoned fields are the habitat for several species of clubmoss, including common running-pine (*Lycopodium digitatum*), tree clubmoss (*L. obscurum* var. *obscurum*), and ground-cedar (*L. tristachyum*), and of grape fern, including cut-leaved (*Botrychium dissectum*) and leathery grape fern (*B. multifidum*). Ebony spleenwort (*Asplenium platyneuron*) and Christmas fern occur in older successional woods and persist into more mature forest.

The Ridge and Valley Province is the richest in number of species and hybrids of ferns and fern-allies in the state. Sussex County has herbarium records for 94 taxa of pteridophytes (Montgomery and Fairbrothers 1985); few counties in the United States exceed this number of pteridophyte taxa. This number represents nearly a quarter of the pteridophytes of the continental United States. Five species of pteridophytes are restricted in their distribution in New Jersey to the Ridge and Valley Province, and six additional species occur only in the Ridge and Valley Province and Highlands Province (Table 2). Most of these are species of northern distribution that reach or approach their southern limit in the state (see below).

The Highlands

The Highlands Province borders the Ridge and Valley on the southeast and includes portions of Warren, Sussex, Hunterdon, Somerset, Morris, Passaic, and Bergen counties (Map 1). The Highlands are underlain by Precambrian rocks that have been highly deformed during the nearly 1 billion years since their deposition into schist, gneiss, and marble. In general the Highlands form a belt of flat-topped ranges trending northeast to southwest separated by deep, narrow valleys of younger rocks (Wolfe 1977). The northern part of the Highlands, like the Ridge and Valley, has been modified by glaciation.

The soils of the Highlands are brown podzolic sandy loams with pH of 4.6–5.1 on the uplands. Gleyed soils and peat deposits occur in depressions (Tedrow 1986). South of the glacial moraine the soils are older red-yellow podzolics, with pH around 4.5.

Vegetation in the Highlands in very similar to that of the Ridge and Valley Province and is similarly controlled by slope and exposure. Ridge tops are occupied by chestnut oak forest; lower slopes typically have a mixed oak forest or eastern hemlock–hardwood forest. Swamps, bogs, marshes, and ponds also occur in the northern, glaciated section and have vegetation similar to that of the Ridge and Valley. The pteridophytes are also similar in both provinces. Two species of pteridophytes are restricted to the Highlands Province in New Jersey (Table 2). The lack of rich woodlands contributes to the lower number of species in the Highlands, and limestone outcrops are much less frequent. Swamps and bogs do occur, and have species similar to these habitats of the Ridge and Valley.

The Piedmont Lowland

The Piedmont Lowland forms a broad belt south and east of the Highlands, extending from Bergen County in the northeast to Mercer County in the west. The greatest density of population, highways, industry, and pollution in New Jersey occurs in this province, and little remains of natural diversity.

The Piedmont Lowland is underlain by rocks of late Triassic to Jurassic ages, consisting of brown to red-brown sandstone and mudstone (Stockton), black to red-brown sandstone, shale, and argillite (Lockatong), and soft red shale (Brunswick) (Wolfe 1977). The Piedmont Lowlands form a gently rolling plain broken by distinct ridges and hills of volcanic origin, including the three ranges of the Watchung Mountains, the Palisades, Rocky Hill, and Cushetunk Mountain. Snake Hill and Little Snake Hill, in the Hackensack Meadowlands, were small plugs of volcanos that extended above the Palisades; they have been nearly eradicated in the course of commercial bluestone production.

Soils of the glaciated uplands are brown podzolic stony to gravelly loams, with pH of approximately 5.0. South of the terminal moraine the soils are derived from the shale and sandstone and tend to be shallow with distinctive red color derived from unaltered rock particles. Drainage is poor and pH is acidic, about 4.5–5.0. Soils derived from the Stockton gray sandstone have red-yellow podzolic features and are heavy clay soils, whereas those derived from the Lockatong argillite are very poorly drained hydric silt loams (Tedrow 1986).

Vegetation of the Piedmont is similar to that of the Ridge and Valley; however, there is more flat land and ridges are not as pronounced. Chestnut oak and red oak are the principal tree species on the ridges. Vegetation of the cliffs of the Palisades and Watchungs is similar to that of cliffs in the Ridge and Valley. Lower slopes and well-drained valleys on the acid soil developed from red shale usually have a mixed oak forest. Dogwood is the most common understory tree, and shrubs are common. There are many spring-flowering herbs and fall-flowering asters (*Aster* spp.) and goldenrods (*Solidago* spp.). Hemlock-dominated forests occur in the ravines of the Watchungs and Palisades (Robichaud and Buell 1973).

Pteridophytes are fewer in the Piedmont Lowland, partly because the relatively flat land provides fewer sheltered moist habitats and fewer species are adapted to the heavy clay soils of the red shale. In the richer forests of the Watchungs, species are similar to those of the richer oak forests of the Ridge and Valley: Christmas fern, northern lady fern, marginal wood fern, and interrupted fern. The cliffs of the Palisades and Watchungs provided habitats for several pteridophytes, but most of these have vanished because of habitat destruction and possibly increased air pollution (Montgomery 1989). Snake Hill, in Hudson County, has herbarium records for 13 taxa, including rock spikemoss (*Selaginella rupestris*) and hairy lip fern (*Cheilanthes lanosa*), both now rare in the state (Montgomery 1982a). All of these species have disappeared, along with the destruction of most of Snake Hill.

Swamp forests occur in the glaciated northern part of the Piedmont. As in the Ridge and Valley, these are often dominated by red maple. Swamps usually have several fern species, including royal fern, cinnamon fern, sensitive fern, crested wood fern, and marsh fern.

Two species and five hybrids of ferns were restricted to the Piedmont Lowland of New Jersey; all of these are now extirpated from the state flora (Snyder 1988; Montgomery 1989). Slender rock-brake (*Cryptogramma stelleri*) and log fern (*Dryopteris celsa*), along with five hybrids of log fern, were found in Bergen County. It is significant that seven of the nine taxa for which there are no recent New Jersey records (since 1950) are from the Piedmont.

The Coastal Plain

The Coastal Plain occupies the southern three-fifths of New Jersey. It is a broad, nearly flat plain, sloping gently eastward to the Atlantic Ocean. The Inner Coastal Plain is older geologically, underlain by sediments of Cretaceous age, consisting of unconsolidated sand, silt, clay, and greensand or marl (Wolfe 1977). The Outer Coastal Plain is underlain by sediments of Tertiary age and also consists of unconsolidated sediments deposited when the Coastal Plain was submerged (Wolfe 1977). The border between these divisions of the Coastal Plain is in some dispute, and the vegetation types do not closely coincide with the geological boundary (Tedrow 1986).

Stone (1911) listed five "floral districts" on the Coastal Plain of southern New Jersey: the Middle District, the Pine Barrens, the Coastal Strip, the Cape May District, and the Maritime District. These districts follow the vegetation pattern closely and thus are used in our discussion of the distribution of New Jersey pteridophytes on the Coastal Plain (the Maritime District has few pteridophytes and is not discussed here).

The Middle District includes the Inner Coastal Plain as well as a part of the Outer Coastal Plain west of the Pine Barrens District (Map 2). The chief soils are yellow sand and gravel, developed from yellow sands, and brown to olive-brown sandy loams, developed from the greensand (Tedrow 1986).

The original flora of the Middle District has been largely exterminated by agriculture and the heavily traveled "corridor" from New York to Philadelphia. Recently, farming has been increasingly abandoned, and housing developments, shopping malls, and retirement villages now occupy much of the area. The dominant vegetation was an American beech (*Fagus grandifolia*) and oak forest, with an understory of heaths in drier sites and dogwood, ironwood (*Carpinus caroliniana*), and sassafras (*Sassafras albidum*) in more mesic sites. In moist-to-wet areas, sweet gum (*Liquidambar styraciflua*) is dominant. Few ferns are characteristic of the Inner Coastal Plain forest; however, Stone (1911) pointed out that climbing fern (*Lygodium palmatum*) was characteristic of this region. More than 15 stations are known to have existed historically for climbing fern from the Inner Coastal Plain; nearly all are gone. Northern lady fern and Christmas fern are found in moist oak or beech woods, and cut-leaved grape fern is found in moist young hardwood stands. Common running-pine and ebony spleenwort are found in open second growth stands in the Middle District. Sensitive fern, cinnamon fern, and marsh fern occur in lowland forests.

Stone (1911) indicated that the Middle District flora extends onto the Outer Coastal Plain into Cape May County and additional southern species are added to the flora. The forests are of southern oaks with pitch pine and American holly (*Ilex opaca*). Lowland forests also include southern pines and oaks. Ferns are not common in the upland woods but are abundant in the lowland forests: Virginia chain fern (*Woodwardia virginica*), cinnamon fern, and marsh fern are often the dominant herbaceous vegetation. Black-footed quillwort (*Isoetes melanopoda*) is restricted to clay "bogs" of Cape May County, disjunct from its main range in the Midwestern states.

The Coastal Strip lies east of the Pinelands and west of the salt marshes and estuaries of the Maritime District. The flora is a mixture of vegetation from the Pine Barrens and Middle Districts (Stone 1911), with upland forests of oak and pitch pine. Northern bracken fern is common in upland woods, and Massachusetts fern is found in the freshwater edges of the coastal marshes.

The Pinelands

The Pinelands occupy a central region of the Outer Coastal Plain and are certainly the most botanically famous region of New Jersey. The New Jersey Pinelands are a distinct natural region, encompassing more than a million acres (more than one-fourth of the state). The approximate boundary of the Pinelands is shown in Map 2. The Pinelands are the single largest tract of land with very low population density and industrial usage in the northeast urban corridor. The Pinelands have commonly been called the Pine Barrens. The term "barrens" originated in the seventeenth century because the area was agriculturally uneconomical. The region was also known as the Pines or Pinelands of New Jersey and is far from barren in terms of natural diversity.

The National Parks and Recreation Act of 1978 created the nation's first natural reserve: the Pinelands National Reserve. The Pinelands Protection Act was enacted by the New Jersey Legislature in June 1979, creating the Pinelands Commission as the regional planning and management agency for the Pinelands National Reserve. The Pinelands Commission established the official name Pinelands for the region after considering several alternative names (Collins and Russell 1988).

The Pinelands are largely underlain by the Cohansey Sand. This is a beach and gravel deposit laid down along a barrier island or beach and tidal flat. The soils in the Pinelands are excessively drained sandy gray podsols, which are strongly acidic in pH and impoverished in nutrients (Tedrow 1986). The lowlands have saturated podsols with very acid pH.

The vegetation and flora of the Pinelands have been the subject of many publications. Detailed descriptions and discussion of various aspects of the ecology of this region can be found in Stone (1911), Harshberger (1916), Robichaud and Buell (1973), McCormick (1979); see additional papers in Forman (1979) and references cited in Buchholz and Good (1982). The informal account of McPhee (1968) is highly recommended for an overview of the history and culture of the people living in this area.

Repeated cutting and severe fires have shaped the vegetation of the Pinelands. Most of the upland has been burned at intervals of 10–30 years, resulting in a unique fire-adapted vegetation (McCormick 1979). The lowland vegetation, while not as subject to fire, has been extensively disturbed by logging and agriculture.

Upland forests of the Pinelands are dominated by various mixtures of pitch pine, shortleaf pine (*Pinus echinata*), black oak (*Quercus velutina*), white oak (*Q. alba*), chestnut oak (*Q. prinus*), post oak (*Q. stellata*), blackjack oak (*Q. marilandica*), and scrub oak (*Q. ilicifolia*) (McCormick 1979). The overall aspect is an open forest of small trees with loose crowns, a dense shrub layer, and much light reaching the forest floor, which is covered with leaf and branch litter rather than herbs. Bracken fern is frequently present and may form a nearly continuous cover, especially after repeated fires. The wide-creeping underground rhizome of this fern usually survives fire, and new fronds emerge within a few days. Southeastern bracken (*Pteridium aquilinum* var. *pseudocaudatum*) also occurs in the Pinelands.

Lowlands occupy 15–20% of the Pinelands, occurring in sites where the water table is at or near the ground surface for some part of the year (McCormick 1979). Atlantic white cedar (*Chamaecyparis thyoides*) occurs in dense, nearly pure stands along many of the streams of the Pinelands, forming dark green ribbons among the gray-green upland pine forests as seen from the air. The shrub and herb layers may be almost absent in dense white cedar stands because of the low light intensity. Cinnamon fern and Virginia chain fern are occasionally found in Atlantic white cedar forests; however, they are more frequent in hardwood swamps. Curly-grass fern (*Schizaea pusilla*) is also found in young or more open Atlantic white cedar stands, although it is more abundant in open bogs.

Red maple is dominant in hardwood swamps of the Pinelands; black gum (*Nyssa sylvatica*) and sweet bay (*Magnolia virginiana*) are associates. Shrubs are more dense than in Atlantic white cedar forests. Cinnamon fern, Virginia chain fern, netted chain fern (*Woodwardia areolata*), and royal fern are typical ferns in hardwood swamps. Along some of the streams in the Pinelands open bogs have developed. Shrubs such as leatherleaf (*Chamaedaphne calyculata*), sheep laurel (*Kalmia angustifolia*), and cranberry (*Vaccinium macrocarpon*) occur in clumps along with young plants of Atlantic white cedar. The dominant vegetation is herbaceous, and includes many of the plants for which the Pinelands are famous. This is the typical habitat for the curly-grass fern (q.v.), along with foxtail clubmoss (*Lycopodium alopecuroides*), appressed bog clubmoss (*L. appressum*), Carolina clubmoss (*L. carolinianum*), and hybrid bog clubmoss (*L.* × *copelandii*). Curly-grass fern and Carolina clubmoss are restricted to the Pinelands in New Jersey.

Distribution Patterns

Thirteen species of pteridophytes are found in all 21 of New Jersey's counties, and six others are recorded from all counties except one (Table 3). All of these species are widespread in the eastern deciduous

Table 3. Pteridophytes Known from All or All Except One County in New Jersey

All counties
Equisetum arvense
Botrychium dissectum
Botrychium virginianum
Osmunda cinnamomea
Osmunda regalis
Polypodium virginianum
Dennstaedtia punctilobula
Thelypteris noveboracensis
Thelypteris palustris
Asplenium platyneuron
Onoclea sensibilis
Athyrium filix-femina var. *angustum*
Dryopteris carthusiana
All except one county[a]
Lycopodium digitatum (Hudson)
Adiantum pedatum (Cumberland)
Pteridium aquilinum var. *latiusculum* (Hudson)
Polystichum acrostichoides (Hudson)
Dryopteris marginalis (Cape May)
Dryopteris intermedia (Middlesex)

[a]Exceptions are shown in parentheses.

forest. Hybrids are excluded in the discussion that follows, for the most part, since their occurrence depends upon the presence of the parents. No hybrid pteridophyte is found in New Jersey without both parents also present, although *Equisetum* × *trachyodon* was discovered before *E. variegatum* was known in the state.

Four species are known from a single county in New Jersey. Two of these are known from a single station: *Isoetes lacustris* (Sussex) and *Cryptogramma stelleri* (Bergen). The other two are known from two or three stations in one county: *Isoetes melanopoda* (Cape May) and *Dryopteris celsa* (Bergen). The two species from Bergen County are now extirpated in the state, and these are the only species not recorded since 1950 [*Marsilea quadrifolia,* considered extirpated by Montgomery and Fairbrothers (1985) has been rediscovered in Sussex County].

Both of the other species known from a single county are considered endangered in New Jersey. Montgomery (1989) listed twelve species as endangered in the state, one as possibly threatened, and four as rare. Snyder (1988) listed ten species as critically imperiled in the state although secure globally, two as critically imperiled in the state and rare and local throughout their range, five as imperiled in the state although secure globally, and six as rare in the state and secure globally. Details are discussed with each species. The extirpated and rare pteridophytes

Table 4. Pteridophyte Taxa Reaching or Approaching Geographic Range Limits in New Jersey[a]

Lycopodium alopecuroides	n	(se MA)
Lycopodium annotinum	s	(sw VA)
Lycopodium carolinianum	n	(se NY)
Lycopodium inundatum	s	(sw VA)
Isoetes echinospora	S	
Isoetes lacustris	s	(w VA)
Isoetes melanopoda	E	
Isoetes tuckermannii	S	
Equisetum pratense	S	
Equisetum variegatum	S	
Botrychium simplex	s	(sw VA)
Botrychium multifidum	s	(w VA)
Ophioglossum vulgatum	N	
Schizaea pusilla	S	
Cryptogramma stelleri	S	
Cheilanthes lanosa	n	(CT)
Pteridium aquilinum var. *pseudocaudatum*	n	(se MA)
Asplenium bradleyi	n	(s NY)
Asplenium montanum	n	(w MA)
Asplenium pinnatifidum	N	
Matteuccia struthiopteris	s	(n VA)
Athyrium filix-femina var. *asplenioides*	n	(MA)
Gymnocarpium dryopteris	s	(VA)
Cystopteris protrusa	E	
Dryopteris clintoniana	S	

[a]N = reaches northern limit in New Jersey; n = approaches northern limit in New Jersey (limit in parentheses); S = southern, E = eastern, etc.

are all near their geographic range limits in the state. Curly-grass fern is the only species designated rare on a national scale (see the discussion under this species).

Twenty-five taxa (species or varieties) of pteridophytes are at or near geographic range limits in New Jersey (Table 4). Two species reach their northern limit here, and seven are near their northern limit, with ranges extending north to southern New York or southern New England. Seven species reach their southern limit in the state (in eastern North America), and six are near their southern limit, with ranges extending south, usually in the mountains, to Virginia. Two species whose ranges are predominately Midwestern reach their eastern limit in New Jersey.

No species of pteridophyte is restricted in its geographic distribution to New Jersey. In the United States, curly-grass fern occurs only in New Jersey except for a small population on Long Island, but this species also occurs in eastern Canada.

How to Use This Book

The following format is used in presenting the treatment of ferns and fern-allies in this book. Each genus is briefly described; if more than a single species occurs in New Jersey, a taxonomic key to the species is provided. All species are included in the key, in addition to varieties and hybrids if more than five stations (localities) are known in New Jersey or they are of special interest (for cultivation or historical information). Varieties and hybrids that do not meet these criteria are discussed either in a separate paragraph or within the parent species treatment; they are usually not illustrated or provided with a distribution map. The scientific name of the species, variety, or hybrid is given first, followed by synonyms that may cause confusion from other fern references. A common name follows; common names of ferns are a matter of choice and are often regional or local. We have chosen one name, but it should be understood that readers will encounter others and that no one name is more correct than any other; this provides the best argument for the use of scientific names by naturalists. [A list of common names of ferns and fern-allies of the United States was compiled by Weatherby (1952) and may be used for purposes of comparison.] The treatment for each species, variety, and hybrid begins with a *Description* that is sufficient to identify the taxon; however, this is not intended to be a complete morphological treatment. For such complete information see Fernald (1950) or Lellinger (1985). The *Habitat* is described, followed by the geographic *Range.* Abbreviations used are the standard two-letter postal code for states and Canadian provinces, and letters for direction modifiers, i.e., n = northern, ne = northeastern, etc. *Distribution in New Jersey* is described briefly and plotted on a state *map.* The maps indicate all known locations in the state (except those too close together to map separately) for the species. Three time periods are indicated by the size of the location dot; this will indicate the time period for the most recent collection from that location. A small dot (·) indicates collections only prior to 1900; a medium-size dot (•), collections 1900–1950; and a large dot (●), collections since 1950. These date choices correspond to the history of fern collecting in New Jersey. Prior to 1900 most of the collecting in the state was by botanists from Philadelphia and New York interested in regional floras, and much of New Jersey was also rural before the beginning of the twentieth century. Between 1900 and 1947, there was rapid urbanization in the areas near New York and Philadelphia, and at the same time an increase in interest in the flora of New Jersey. This time period included the publication of two major floristic works: Stone's "The plants of southern New Jersey . . ." (1911) and Chrysler and Edwards's *The Ferns of New Jersey* (1947). The period from 1950 to 1990 is considered "recent" for purposes of this book, which includes collections up to 1990. Additional details about the botanists who collected in New Jersey—and thus made this book possible—are included in the Historical Introduction.

The section *Comments* is used for a variety of information about the species, including special notes about distribution, early records, status if rare or endangered in New Jersey, subspecies other than those occurring in New Jersey, recent information about the taxonomy of the species or group, related taxa that have been found in bordering states but not in New Jersey (and therefore might be sought in the state), cultivation information, and the chromosome number. Chromosome numbers are included since they have proven of significant value in understanding fern taxonomy, especially in hybrid complexes. The numbers are those published by Löve et al. (1977) unless otherwise noted.

Each species is illustrated with a line drawing of the entire plant or a frond, together with details of critical features usually including the sorus and indusium. These illustrations supplement the description and aid in identification.

This format is followed for all species of pteridophytes that occur or have occurred in New Jersey based on vouchered herbarium records.

Fern-Allies

Lycopodium

Plants with short-to-long creeping rhizomes, these producing erect branches, with needle-like or scale-like leaves with only a midvein; sporangia in axils of leaves (sporophylls); sporophylls like the vegetative leaves and in zones along the stem, or more or less modified and grouped into strobili at the ends of the erect stems or branches; spores all alike.

Lycopodium is a large genus (400–500 species), worldwide in distribution. Many species are epiphytes in tropical forests; however the genus is well represented in temperate and boreal regions. This is an ancient group of plants, with relatives of tree size in the coal-forming forests of the Paleozoic Era.

Lycopodium has included, since the time of Linnaeus, divergent taxonomic elements (Linnaeus included *Selaginella* and *Psilotum*). Several efforts have been made to divide the genus *Lycopodium,* in its modern sense, into subgenera or genera (Boivin 1950). Recent authors have recognized at least four genera (Øllgaard 1987) and up to a dozen genera placed in two families (Holub 1985). The evidence for division of this group into four genera is based on extensive differences in branching patterns (Øllgaard 1979), anatomy, presence or absence of distinct strobili and modified sporophylls, spores (Wilce 1972), and gametophytes (Øllgaard 1987). The remaining splinter genera comprise one or a few species each differing in only one or two characteristics; they are thus best taxonomically treated as subgenera.

In this book the traditional treatment (single genus) has been followed in both the key and descriptions because most users will be familiar with this arrangement. The generic and specific epithets of Øllgaard's classification are indicated in the key and species descriptions, since we believe these names should be accepted for a modern treatment of this group. The key and species treatments are based on Beitel's (1979) excellent review of this genus in North America.

A note is in order concerning the confusing common names applied to this group of pteridophytes. Almost any combination of clubmoss, pine, fir, or cedar, with the modifier common, ground, running, trailing, or princess has been used for one or more species in the subgenus *Lycopodium;* and Weatherby (1952) listed 24 common names from the literature for *L. clavatum* alone! We have followed the choices from Beitel's treatment (1979).

Except for *L. lucidulum,* which can be grown in a woodland garden, most species are difficult to transplant, and they cannot readily be grown from spores.

Key to Subgenera (Genera), Species, and Hybrids

1. Stems unequally branched from long-creeping rhizomes; sporophylls modified and aggregated into distinct strobili **2**

2. Fertile erect branches branching, tree-like or antler-like in form. Subgenus *Lycopodium* (*Lycopodium*) **3**

3. Leaves spirally arranged, in 6–10 rows, not fused to stem **4**

4. Upright stem with few branches; plants antler-like in form; leaves in 8–10 rows; rhizome near soil surface **5**

5. Strobili on long peduncles; leaves in ten rows, with a bristle tip, not serrate.................................. *L. clavatum*

5. Strobili sessile; leaves in 8 rows, without bristle tip, serrate, ... *L. annotinum*

4. Upright stem with several regular branches; plants tree-like in form; leaves in 6 rows; rhizome buried in soil **6**

6. Leaves on lower part of upright stem (before first branch) appressed to stem; leaves arranged with one row on top of branch, two on each side, and one on bottom .... *L. obscurum*

6. Leaves on lower part of upright stem spreading; leaves arranged with a double row on top and bottom of branch and one on each side (see illustration).......... *L. dendroideum*

3. Leaves in pairs arranged in 4 rows; partly fused at base to the stem **7**

7. Rhizomes near or on soil surface; free tips of lower leaves one-fourth as long as tips of side leaves; branches without annual constrictions..................................... *L. digitatum*

7. Rhizomes buried in the soil at least in part; free tips of lower leaves one-half or longer than tips of side leaves; branches with at least some annual constrictions **8**

8. Branch constrictions occurring irregularly; rhizome depth variable...................................... *L.* × *habereri*

8. Branch constrictions regularly present; rhizomes deeply buried in soil.................................. *L. tristachyum*

2. Fertile erect branches mostly simple, unbranched: Subgenus *Lepidotis* (*Lycopodiella*) **9**

9. Rhizomes with side leaves much longer and wider than top leaves; strobili with modified sporophylls, peduncles nearly leafless .... *L. carolinianum*

9. Rhizomes with all leaves alike; strobili with sporophylls only slightly modified from vegetative leaves; peduncles leafy............ **10**

10. Rhizomes 0.6–1.2 mm thick (excluding leaves); strobili accounting for one-third to one-half the height of erect stems; erect stems 2–8 cm high .... *L. inundatum*

10. Rhizomes 1–4 mm thick (excluding leaves); strobili accounting for less than one-fourth of height of erect stems; erect stems 5–30 cm long.. **11**

11. Strobili 15–20 mm thick, with spreading sporophylls; rhizomes arching above the substrate, rooting only at the ends; leaves of erect stems spreading, with hair-like teeth along the margins .... *L. alopecuroides*

11. Strobili 3–12 mm thick, with ascending-to-appressed sporophylls; rhizomes creeping or somewhat arching; leaves of erect stem ascending to appressed, with 0–2 teeth per side **12**

12. Stem leaves and sporophylls ascending; with 1 or 2 teeth per side; strobili 5–10 mm thick............ .... *L. × copelandii*

12. Stem leaves and sporophylls appressed; entire; strobili 3–5 mm thick..*L. appressum*

1. Stems with equal branching; rhizome short-creeping to nearly erect; sporophylls and vegetative leaves alike, strobili lacking: Subgenus *Urostachya* (*Huperzia*)... *L. lucidulum*

Lycopodium clavatum L.

Staghorn Clubmoss

Description: Rhizomes creeping in litter, producing forked upright branches; leaves in ten rows, making the branches cylindrical, all alike, nearly linear, each tipped with a hair-like bristle; strobili several on long peduncles.

Habitat: Dry to moist acidic rocky woods and old fields; often on drier ridges in swamps.

Range: Greenland to AK, south to mountains of NC, KY, OH, MN, MT, and n CA; also Central America and Eurasia.

Distribution in New Jersey: Scattered in the northern counties and on the Inner Coastal Plain.

Comments: *Lycopodium* spores, which can be collected in abundance from species that produce strobili, have been used for fireworks, chemistry, and medicine. They contain an oil that is not only highly flammable but makes the spores water resistant, so they can be used like talcum powder in dressings. Balick and Beitel (1989) reported that the spores can cause allergic reactions and noncancerous granulomas when used as lubricants in condoms.

Material from New Jersey is var. *clavatum;* another variety occurs farther north in North America. The chromosome number of *L. clavatum* is $2n = 68$.

Lycopodium clavatum L.

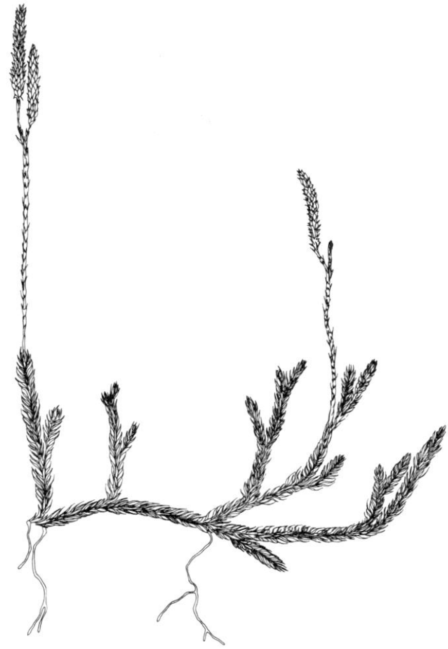

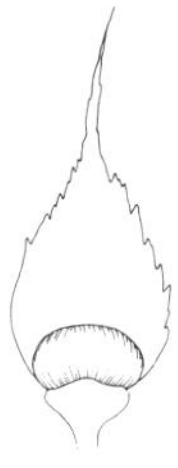

Lycopodium annotinum L.

Bristly Clubmoss

Description: Rhizomes creeping near soil surface or in litter; erect stems simple or few-forked; leaves in eight rows, narrowly elliptic, serrate; strobili sessile, solitary at the branch ends.

Habitat: Acidic, often conifer woods and rocky open areas.

Range: Greenland to AK, south to New England, n NJ, mountains of sw VA, and MI to MN and OR; also Eurasia.

Distribution in New Jersey: One station in Sussex County, and an older record from Morris County.

Comments: Both stations were discovered by J. L. Edwards in 1933. This species, near the southern limit of geographic distribution in New Jersey, was listed as endangered in the state by Montgomery (1989), rare by Fairbrothers and Hough (1973), and critically imperiled in the state, although globally secure, by the New Jersey Natural Heritage Program (Snyder 1988).

L. annotinum looks superficially like *L. lucidulum* when not fertile, but the rhizomes are wide-creeping and all leaves are the same length; the two species often grow together on the Pocono Plateau, forming large evergreen patches in the upland woods.

The chromosome number of *L. annotinum* is $2n = 68$.

Lycopodium annotinum L.

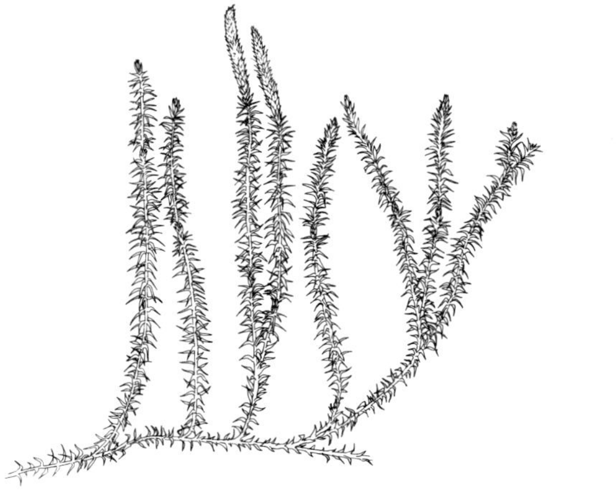

Lycopodium obscurum L.

Tree Clubmoss

Description Rhizomes creeping, buried in soil; erect stems branched, tree-like; leaves linear, in six ranks, with top and bottom rows and four lateral rows in end view (like the even numbers on a clock); leaves of erect stem below first branch appressed to stem; strobili sessile.

Habitat: Damp to sometimes dry open woods and thickets, and drier parts of swampy woods; moist oak-pine woods in s NJ.

Range: NS to MN, south to SC and AL.

Distribution in New Jersey: Throughout the state, with records from all counties except Hudson and Union.

Comments: In a recent revision of this group, Hickey (1977) clearly distinguished *L. dendroideum* from *L. obscurum*. He also distinguished two varieties of *L. obscurum*. Variety *obscurum* has the lower leaves shorter, and the upper and lower leaves appressed to the branch, giving a flattened appearance. Variety *isophyllum* Hickey has all leaves equal in length and equally divergent from the branches. These are not varieties in the traditional geographic sense, since they occupy almost identical ranges in the eastern United States, and sometimes grow together, where they look quite distinct. Wagner et al. (1989) consider var. *isophyllum* to be a distinct species, which they have named *L. hickeyi*. Unfortunately, the usual biosystematic tests are difficult to perform in *Lycopodium,* since the gametophytes grow underground and take years to mature. The plants are also nearly impossible to transplant. Chemical techniques may aid in determining whether var. *isophyllum* is a growth form, a widespread morphological variety, or a separate species.

The chromosome number of *L. obscurum* is $2n = 68$.

Lycopodium obscurum L.

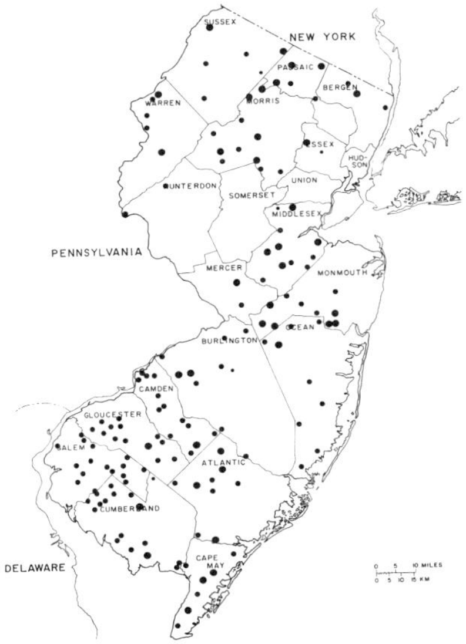

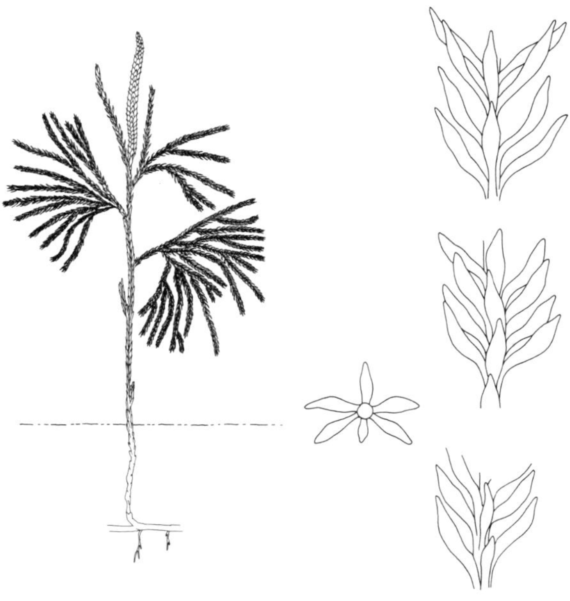

Lycopodium dendroideum Michx.

[*L. obscurum* L. var. *dendroideum* (Michx.) D. C. Eaton]
Tree Clubmoss

Description: Rhizomes creeping, buried in soil; erect stems branched, tree-like; leaves linear, in six ranks, with two rows on top and bottom, and one row on each side in end view (like the odd numbers in a clock); leaves of erect stem below first branch widely spreading; strobili sessile.

Habitat: Moist open woods and thickets.

Range: NF to AK, south to WV, MI, IL, IA, and WA; also in Asia.

Distribution in New Jersey: Widely scattered; most recent records from the Highlands and Piedmont, but older records from all physiographic provinces (and 16 counties).

Comments: *Lycopodium dendroideum* has usually been designated a variety of *L. obscurum*. Hickey (1977) has shown that the two are distinct in several morphological characteristics and should be treated as separate species. The best way to distinguish them in the field is to look at the branches end-on. In *L. dendroideum* the leaves appear in rows equivalent to the odd numbers on the face of a clock. This characteristic is not usually helpful for identification of herbarium material, but the spreading leaves at the base of the erect stem are usually evident.

The chromosome number of *L. dendroideum* is $2n = 68$.

Lycopodium dendroideum Michx.

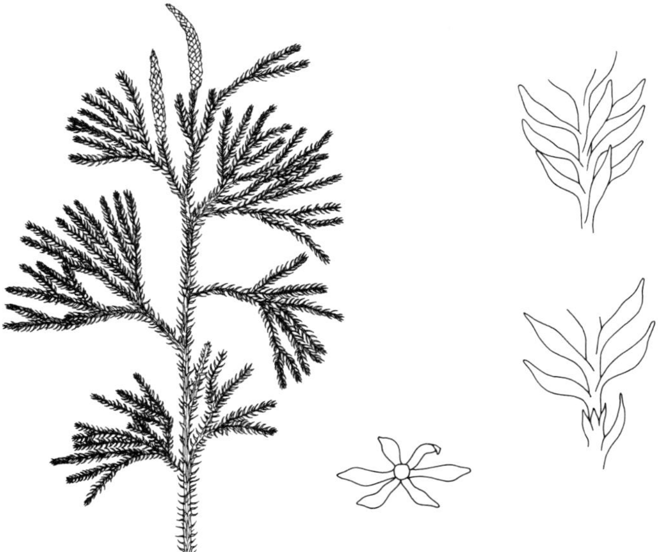

Lycopodium digitatum A. Br.

[*L. flabelliforme* (Fern.) Blanch.]
[*L. complanatum* L. var. *flabelliforme* Fern.]
Common Running-pine

Description: Rhizomes creeping near soil surface or in litter; erect branches regularly branched, lacking annual constrictions; leaves scale-like, fused to the stem for half or more of their length, in four rows, the free tips of the top and bottom leaves much shorter than the lateral leaf tips; strobili on long stalks, the peduncle usually with two branches very close together.

Habitat: Dry to fairly moist open woods, thickets, and abandoned fields; becomes established in "perennial herb" stage of secondary succession and persists through thicket and young woodland stages to more or less open forest.

Range: NF to MN, south to n AL, KY, and IA.

Distribution in New Jersey: Nearly throughout the state but uncommon in the Pinelands; recorded from all counties except Hudson.

Comments: *Lycopodium digitatum* has long been considered a variety of *L. complanatum* but differs from this circumboreal species in several characteristics (Wilce 1965; Beitel 1979). It was called *L. flabelliforme* until Hickey and Beitel (1979) demonstrated that the earlier specific epithet *digitatum* is the correct name.

The colonies of clubmosses in woods are often more or less circular and probably represent clonal growth from a single original plant. These colonies may reach large size and great age. In Finland, Oinonen (1967) observed colonies of *L. complanatum* 250 m in diameter and growth rates of 30 cm/year, giving an age of 830 years.

The chromosome number of *L. digitatum* is $2n = 46$.

Lycopodium digitatum A. Br.

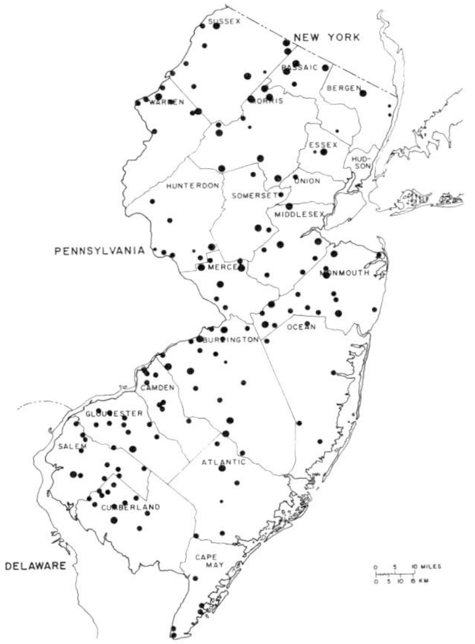

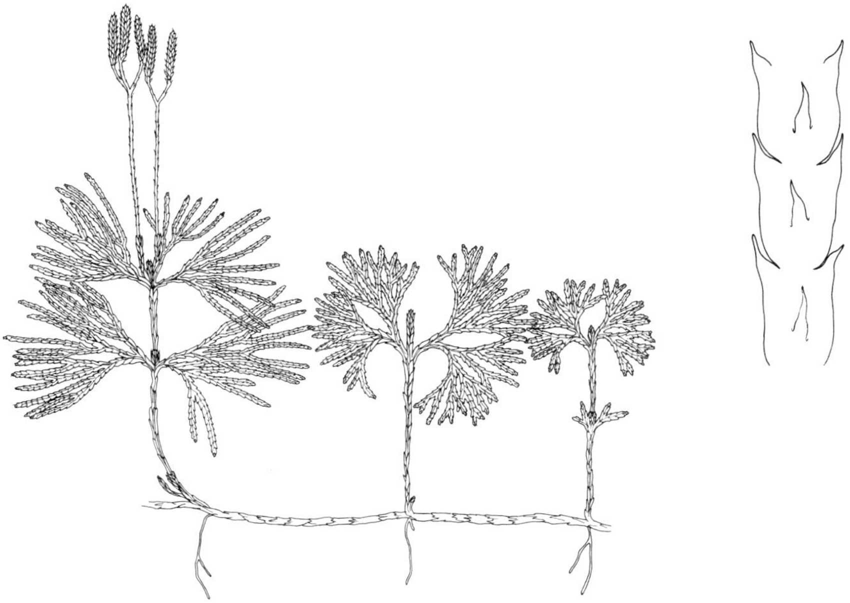

Lycopodium × *habereri* House
L. digitatum × *tristachyum*

Hybrid Running-pine

Description: Rhizomes creeping, partly buried in soil but arching near surface in places; erect branches regularly branched, with occasional annual constrictions; leaves scale-like, fused to the stem for half or more of their length, in four rows, the bottom leaves with the tips more than half as long as the lateral leaf tips; strobili on long stalks, the peduncle branches somewhat separated.

Habitat: Abandoned fields and edges of woods, usually with at least one parent in the vicinity.

Range: Scattered in the area of overlap of the ranges of the parents, from NS to NY and MI, south to VA.

Distribution in New Jersey: Sussex, Morris, and Bergen counties, apparently rare.

Comments: This hybrid, like most hybrids in pteridophytes, is intermediate in morphology between its parents. The rhizome varies in depth from shallow to fairly deep, and there are a few annual branch constrictions. In *L. tristachyum* the branch tips resume growth for a second year, leaving a constriction at the end of the first year's growth; in *L. digitatum* they do not resume growth; and in *L.* × *habereri* they sometimes resume growth. Sun and shade forms of the hybrid and its parents make branch width and color unreliable characteristics, but the comparative length of the leaf tips is reliable.

Wilce (1965) observed that at least some of the spores were abortive.

The chromosome number of *L.* × habereri is $n = 23$.

Lycopodium × *habereri* House

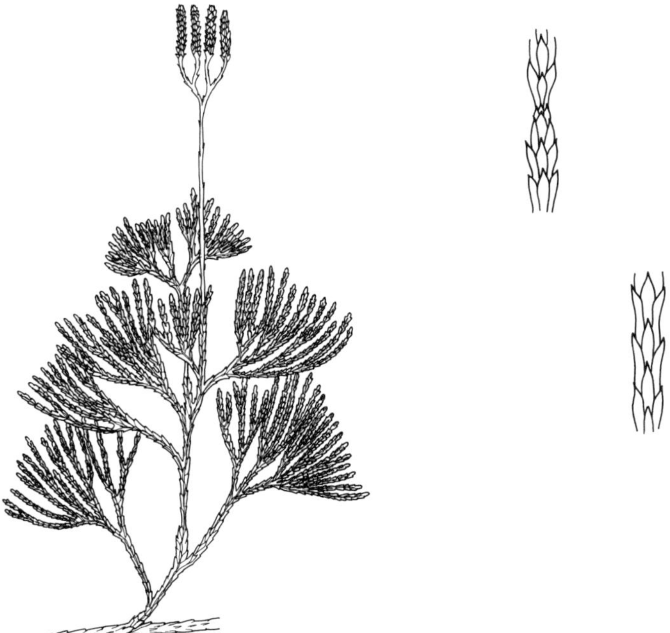

Lycopodium tristachyum Pursh

Ground-Cedar

Description: Rhizomes creeping, deeply buried (5–15 cm) in soil; erect branches more or less regularly branched, with annual constrictions where growth is resumed in second year; leaves scale-like, fused to the stem for up to half their length, in four rows, the free tips of all leaves about the same length; strobili on long stalks, the peduncle branches separated by 0.5 cm or more.

Habitat: Abandoned fields and open second-growth woods, under power lines, and drier upland areas in swamps; unlike *L. digitatum* and *L. obscurum,* this species rarely occurs in woodlands where the canopy is closed.

Range: NF to MN, south to GA, n AL, and MO; also in Europe.

Distribution in New Jersey: Fairly common in the northern tier of counties, absent from the central part of the state, and a few records in s NJ.

Comments: This species is usually not difficult to distinguish from *L. digitatum* by the deeply buried rhizome, the branch constrictions, and the lower leaf tips as long as the lateral leaf tips. Plants of *L. tristachyum* are often blue-green in color, but color varies in sun and shade forms. Wilce (1965) pointed out the difference in the peduncle branches: in *L. tristachyum* the second branch is well above the first; in *L. digitatum* they are very close together. Comparisons with *L.* × *habereri* are included with the hybrid description.

The chromosome number of *L. tristachyum* is $2n = 46$.

Lycopodium tristachyum Pursh

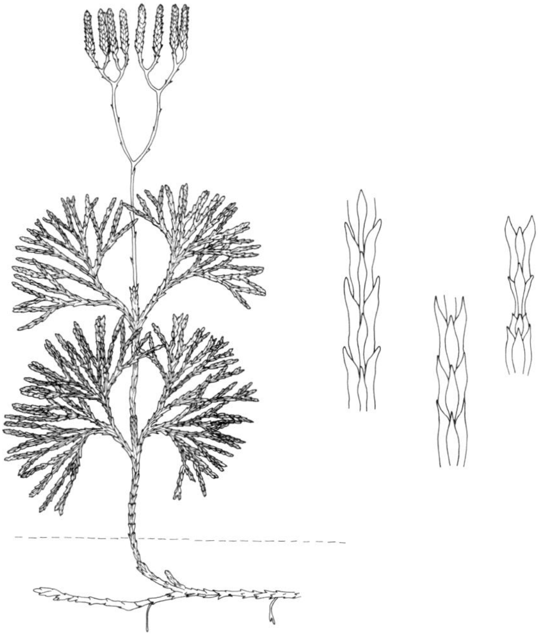

Lycopodium carolinianum L.

Lycopodiella caroliniana (L.) Pichi-Serm.
Carolina Clubmoss

Description: Rhizomes creeping on surface of sand or peat; erect branches unbranched; leaves of rhizome in three to five rows, with lateral rows of larger leaves and upper rows of very small leaves; leaves of erect stems small, in scattered whorls of three (the effect much like a peduncle), with a single strobilus at the summit.

Habitat: Moist sandy or peaty bogs and damp sandy acid depressions such as old roadbeds and edges of cranberry bogs.

Range: Mostly on the Coastal Plain from FL to TX, north to NJ and Long Island, NY; closely related taxa in Central and South America, Africa, Ceylon, Australia, and New Zealand.

Distribution in New Jersey: Fairly common in bogs and wet sandy areas of the Pinelands; in Cape May County formerly in wet bogs underlain with white clay, and at Island Beach formerly in wet swales between sand dunes.

Comments: The morphology and anatomy of *L. carolinianum* was described by Bruce (1976). He reported the chromosome number as $n = 35$ from Georgia and $n = 70$ from Louisiana; triploids were also found.

This species is one of the characteristic plants of the Pinelands. It shares with *Schizaea* not only habitat but difficulty of discovery. Neither species grows in dense *Chamaecyparis* forests, but either may be found in more or less open wet areas where there is little competition from grasses, sedges, and *Sphagnum*. In open bogs look for small cleared areas of damp sand or along the edges of cedar hummocks.

Lycopodium carolinianum L.

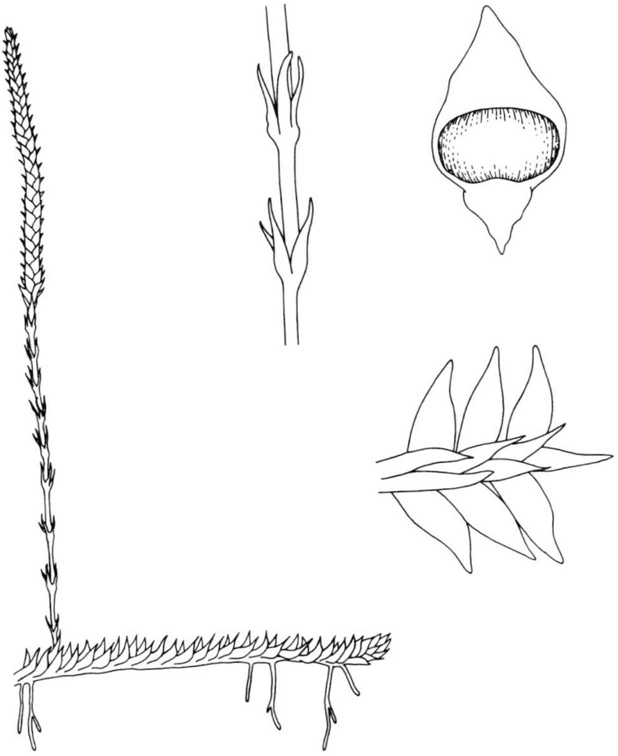

Lycopodium inundatum L.

Lycopodiella inundata (L.) Holub
Bog Clubmoss

Description: Rhizomes creeping on soil or peat surface, rooting throughout, not arching, 0.6–1.2 mm thick; erect branches unbranched, 3–8 cm tall; leaves spirally arranged in many rows, linear, spreading, lacking teeth; strobili up to one-half the total length of erect stems; sporophylls only slightly differentiated.

Habitat: Moist sandy acid soil of roadsides, edges of ponds, and gravel pits.

Range: NF to AK, south to NJ, PA, and mountains to sw VA, west to IL and nw CA; also in n Europe and Asia (circumboreal).

Distribution in New Jersey: Several stations in the northern counties; two old records along the Delaware River.

Comments: This species was considered rare in the state, although globally secure, by the New Jersey Natural Heritage Program (Snyder 1988).

According to Bruce (1975), *L. inundatum* consists of two taxa in the Great Lakes area: true *L. inundatum,* which is diploid, $n = 78$; and *L.* "appressed inundatum," which is tetraploid, $n = 156$. "Appressed inundatum," which does not have a valid name, differs morphologically in having ascending rather than spreading sporophylls, and thick (1.0–2.6 mm) rather than thin (0.6–1.2 mm) rhizomes. Both of these taxa probably occur in northern New Jersey. To make life more difficult, both taxa hybridize with other members of *Lycopodiella* (see *L. inundatum* hybrids).

Lycopodium inundatum L.

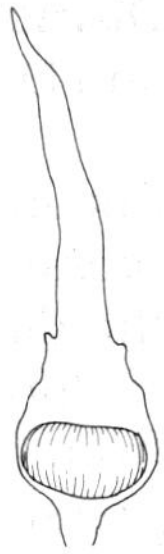

Lycopodium appressum (Chapm.) Lloyd & Underw.

Lycopodiella appressa (Lloyd & Underw.) Cranfill
[*L. inundatum* L. var. *bigelovii* Tuckerm.]
[*L. chapmanii* Underw. ex Maxon in part]
Appressed Bog Clubmoss

Description: Rhizomes creeping on soil or peat surface, rooting throughout, not arching, 1–4 mm thick; erect branches to 30 cm tall, unbranched; leaves spirally arranged in many rows, linear, appressed to the stems, lacking teeth; sporophylls only slightly differentiated at top of the erect stem, appressed; strobili less than one-fourth the total height of erect stems.

Habitat: Moist to sandy wet open depressions and roadsides; open bogs and abandoned cranberry bogs, especially around the edges.

Range: Chiefly on the Coastal Plain, FL to TX, north to NS and NF, also inland in KY.

Distribution in New Jersey: Common in the Pinelands, along the coast and Delaware River, and scattered localities in the northern counties.

Comments: As with *L. inundatum,* Bruce (1975) reported that *L. appressum* consists of diploid and tetraploid taxa: true *L. appressum* is the diploid form ($n = 78$) and occurs primarily on the Coastal Plain; *L.* "northern appressum" is tetraploid ($n = 156$) and was identified from the western Great Lakes area. The latter differs from true *L. appressum* in having a single erect stem 4.5–7.5 cm long per rhizome; *L. appressum* has three to five erect stems 10–30 cm long per rhizome. Both taxa occur in New Jersey; however, Bruce did not study the important area of overlap between the coastal and inland species that occurs in New Jersey and Long Island. More investigation of this group is needed in the Northeast before it is properly understood; Chrysler and Edwards (1947) pointed out this taxonomic problem some 45 years ago!

Lycopodium appressum (Chapm.) Lloyd & Underw.

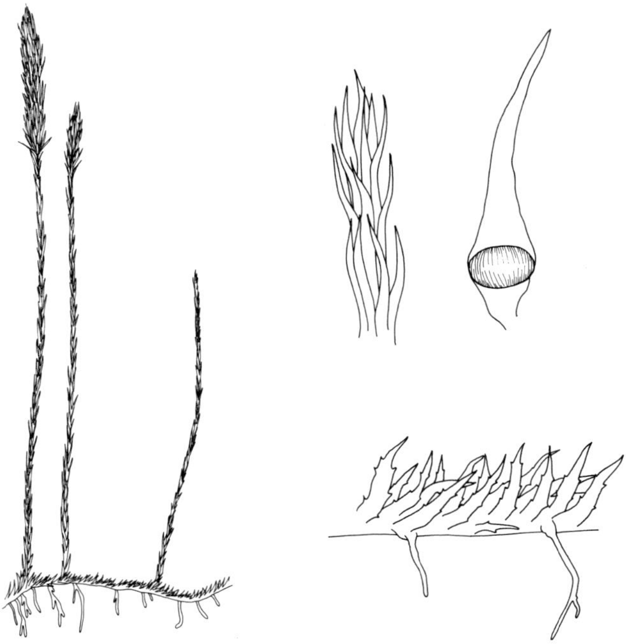

Lycopodium × *copelandii* Eiger
L. alopecuroides × *appressum*

Lycopodiella × *copelandii* (Eiger) Cranfill
[*L. inundatum* L. var. *elongatum* Chapm.]
[*L.* × *chapmanii* Underw. ex Maxon in part]
Hybrid Bog Clubmoss

Description: Rhizomes creeping or somewhat arching over soil, peat, or low vegetation, rooting at ends and sometimes along the surface, 1–4 mm thick; erect branches to 30 cm tall, unbranched; leaves spirally arranged in many rows, linear, more or less ascending to spreading, with one or two pairs of teeth at the base; sporophylls only slightly differentiated at top of erect stems, ascending; strobili less than one-fourth the length of the stems.

Habitat: In wet sandy depressions, open bogs, and roadsides.

Range: Coastal Plain, FL to TX, north to Long Island, NY.

Distribution in New Jersey: Common in the Pinelands; scattered in the northern counties and along the coast.

Comments: Bruce and Evans (1979) pointed out that the name *L. chapmanii,* which has been applied to this hybrid, was published (by W. R. Maxon) as a species name for *L. inundatum* var. *appressum,* which is also called *L. appressum.* Therefore the name *L. chapmanii* is a synonym of *L. appressum* and cannot be used for this hybrid. The name *L.* × *copelandii* is based on material from Lakehurst, NJ. As indicated earlier under *L. appressum,* there are diploid and tetraploid taxa involved in what has been called *L. appressum. L.* × copelandii is apparently the diploid cross; however, Bruce (1975) did not investigate specimens from New Jersey.

Lycopodium × *copelandii* Eiger

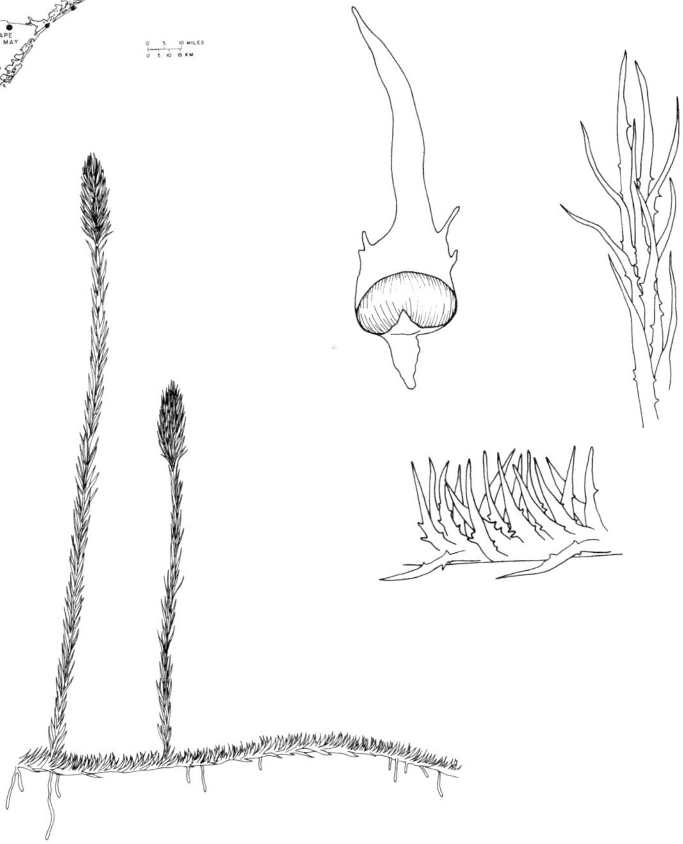

Lycopodium alopecuroides L.

Lycopodiella alopecuroides (L.) Cranfill
Foxtail Clubmoss

Description: Rhizomes arching over substrate, rooting only at the ends, 2.5–3.5 mm thick; erect branches up to 25 cm tall, unbranched; leaves spirally arranged in many rows, linear, widely spreading, toothed at the base; sporophylls only slightly differentiated from vegetative leaves, widely spreading; strobili less than one-fourth the height of the stems.

Habitat: Moist to wet sandy depressions, damp sandy roadsides, and open bogs, including abandoned cranberry bogs.

Range: Fl to TX, north, mostly on the Coastal Plain to Long Island, NY, and Nantucket, MA; also Central and South America.

Distribution in New Jersey: Common in the Pinelands and along the coast; scattered in the northern counties and the Inner Coastal Plain.

Comments: There seems to be no confusion about the identity of this species (Bruce 1975). All material is diploid, with $n = 78$. The arching stems and spreading leaves are distinctive, and serve as good markers in hybrids involving *L. alopecuroides*. Hybrids with *L. inundatum* have the elongated strobili of that species as well as the spreading sporophylls of *L. alopecuroides* and are spectacular small plants with large cones.

It is relatively easy to cultivate this species in a bog terrarium, but plants should be transplanted in early spring, before the bud at the end of the past year's rhizome has begun growth.

Lycopodium alopecuroides L.

Lycopodium (*Lycopodiella*) hybrids involving *L. inundatum*

Description: Rhizomes creeping to the arching, 1–3 mm thick; erect branches 5–20 cm tall, unbranched; leaves linear, usually somewhat spreading, lacking teeth or with one or two pairs of teeth; sporophylls only slightly differentiated, ascending to spreading; strobili one-fourth to one-half the total length of the erect stems.

Habitat: Usually with the parents in damp open depressions, edges of small ponds, and gravel pits.

Range: Uncertain, but in the narrow area of overlap of *L. inundatum* and the coastal species of *Lycopodiella*.

Distribution in New Jersey: One locality in Morris County and a former locality in Sussex County.

Comments: Both *L. inundatum* and *L. appressum* include diploid and tetraploid taxa, and these, as well as *L. alopecuroides*, all occur in the northern Coastal Plain, including New Jersey and Long Island. Hybrids in this group of species were first collected in New Jersey by J. L. Edwards in 1950, and an additional locality with these hybrids has been studied by J. D. Montgomery. The hybrids involving *L. inundatum* (and L. "appressed inundatum") have strobili comprising up to half the erect stem height; those with *L. alopecuroides* have spreading leaves and arching rhizomes; and those of *L. appressum* (and *L.* "northern appressum") have creeping rhizomes and appressed leaves; but the effect when all taxa occur together is baffling. Hybrids involving different chromosome numbers ($2x \times 4x$) produce about 40% aborted spores, but crosses in this group involving taxa with the same chromosome number have mostly normal spores (Bruce 1975). Until further study delimits the taxa geographically and morphologically, such hybrids cannot be identified and named with confidence.

Lycopodium inundatum hybrids

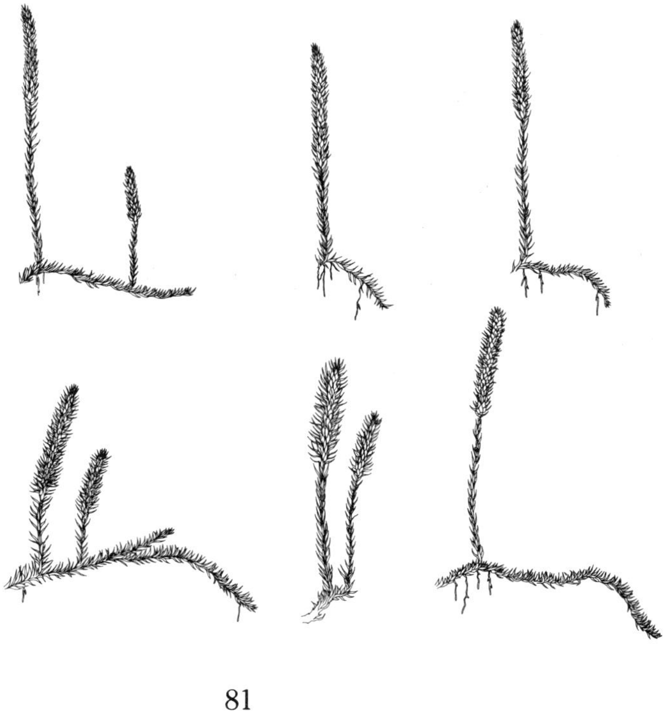

Lycopodium lucidulum Michx.

Huperzia lucidula (Michx.) Travisan
Shining Firmoss

Description: Rhizomes short-creeping, regularly forking to produce erect branches; leaves lanceolate to elliptic, serrate, in many rows; sporophylls in zones along the stem, not forming a strobili.

Habitat: Moist woods, often coniferous, and rocky slopes.

Range: NF to w ON and MN, south to n AL, GA, and AR.

Distribution in New Jersey: Common in the northern counties and formerly on the Inner Coastal Plain.

Comments: Vegetative reproduction occurs by tiny gemmae (three-pointed buds borne near the tip of the stem). The chromosome number is $2n = 67$ (Beitel and Wagner 1982).

Two additional species of this group have been found growing very close to New Jersey across the Delaware River in Pennsylvania. A specimen labeled *Lycopodium selago* L., the fir clubmoss, was once collected on the Pennsylvania side of the Delaware Water Gap by F. W. Knipe in 1870. J. M. Beitel has examined the specimen and considers it to be a hybrid involving *L. selago* and an Appalachian species. *Lycopodium porophilum* Lloyd & Underw. [*Huperzia porophila* (Lloyd & Underw.) Holub], the rock clubmoss, occurs in eastern Pennsylvania at waterfalls along the Pocono Escarpment, but it has never been found in New Jersey, which is only about 2 miles away. *L. porophilum* differs from *L. lucidulum* in having few, if any, teeth on the leaves, and less pronounced zones of longer and shorter leaves on the erect stems (Beitel 1979). The habitat of *L. porophilum* is also very different from that of *L. lucidulum:* the former grows on sandstone cliffs, not on the forest floor. *L. porophyllum* should be searched for in New Jersey.

Lycopodium lucidulum Michx.

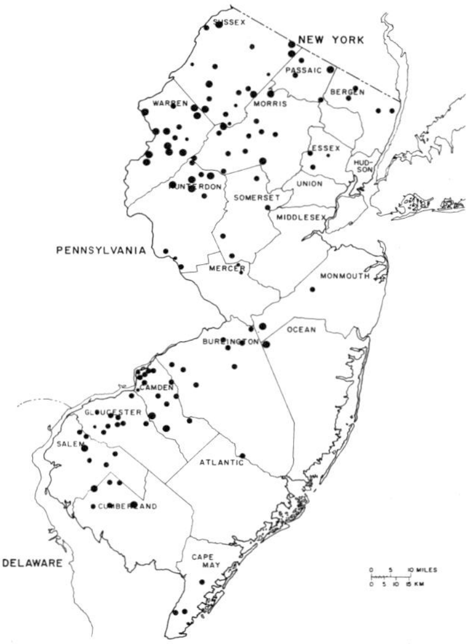

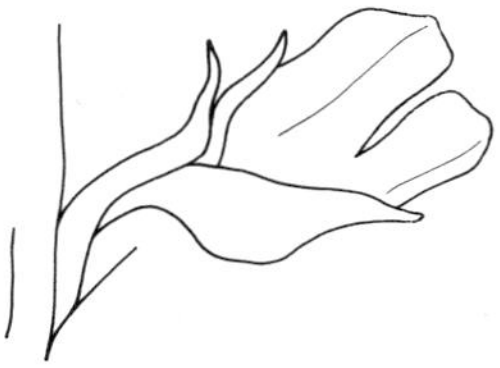

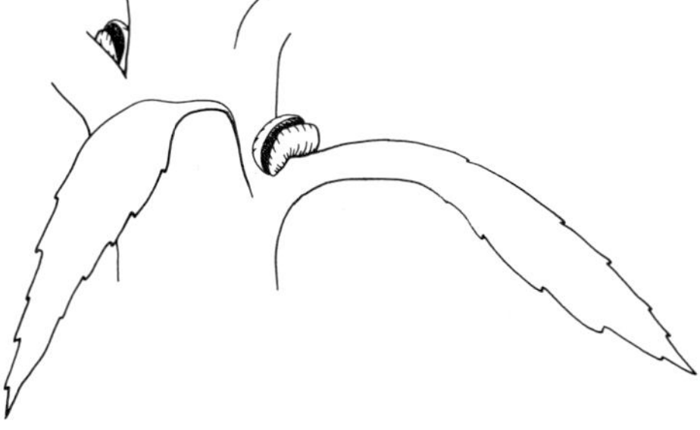

Selaginella

Plants moss-like; stems creeping, with small simple leaves having a midrib only; sporangia in axils of more or less modified leaves (sporophylls), grouped into distinct four-sided strobili; spores of two types (microspores and megaspores).

Selaginella is a large genus (700 species), with two subgenera. Subgenus *Selaginella* has spirally arranged equal leaves and occurs primarily in rocky or sandy habitats, especially in dry areas. Subgenus *Stachygynandrium* has leaves in four rows, with the lateral leaves larger and spreading, and the top and bottom leaves smaller and appressed to the stem; most species are tropical. One species of each subgenus occurs in New Jersey.

Key to Species

1. Leaves spirally arranged, all alike, linear .... *S. rupestris*
1. Leaves in four rows, the lateral spreading, ovate; the top and bottom appressed, acuminate .... *S. apoda*

Selaginella apoda (L.) Spring

Meadow Spikemoss

Description: Stems creeping, forming patches; leaves in four ranks, the lateral spreading, ovate, 1.5–2.0 mm long, the two top rows smaller and appressed, elliptic, 0.8–1.5 mm long.

Habitat: Moist to wet meadows, borders of fresh or brackish marshes, springy slopes, and damp places in lawns.

Range: s ME and PQ to MI, IL, and MO, south to FL, e TX, and OK.

Distribution in New Jersey: Common throughout the state except the Pinelands and the northeastern counties.

Comments: This tiny plant is often overlooked. One of the more common habitats is damp, partially shaded places in lawns, where it goes undetected until the lawn is weeded!

The chromosome number of *S. apoda* is $2n = 18$.

Selaginella apoda (L.) Spring

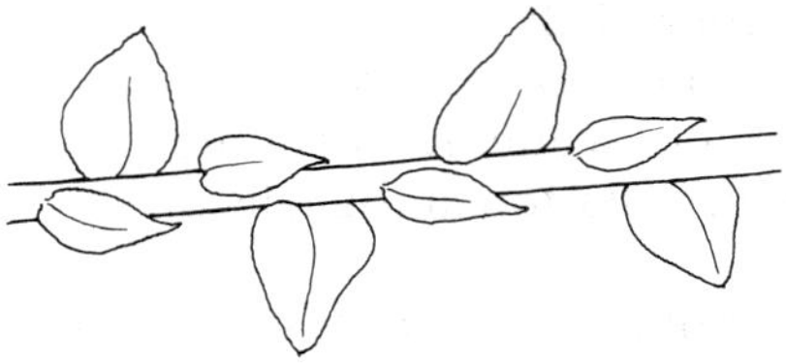

Selaginella rupestris (L.) Spring

Rock Spikemoss

Description: Plants forming mats on rocky ledges or sand; leaves thick-textured, all alike, spirally arranged, linear, 1.0–1.5 mm long, with a bristle tip.

Habitat: On acid rocks thinly covered with soil, or on sandy soil, usually in full or partial sun.

Range: PQ to s MB, south to n NJ, and in the mountains to AL, AR, and OK.

Distribution in New Jersey: Widely scattered in the northern third of the state.

Comments: Most of the localities in the northeastern counties have disappeared, partly owing to urbanization and perhaps air pollution (Montgomery 1989). *S. rupestris* is now restricted to a few sites in the western part of the state. This species was designated as endangered in New Jersey by Snyder and Vivian (1981) and Montgomery (1989) and was considered imperiled in the state by the New Jersey Natural Heritage Program (Snyder 1988).

S. rupestris is one of the early extant herbarium records for New Jersey: a specimen at the Academy of Natural Sciences in Philadelphia was collected by B. S. Barton at the falls of the Passaic (Paterson) in 1806.

The chromosome number of *S. rupestris* is $2n = 18$.

Selaginella rupestris (L.) Spring

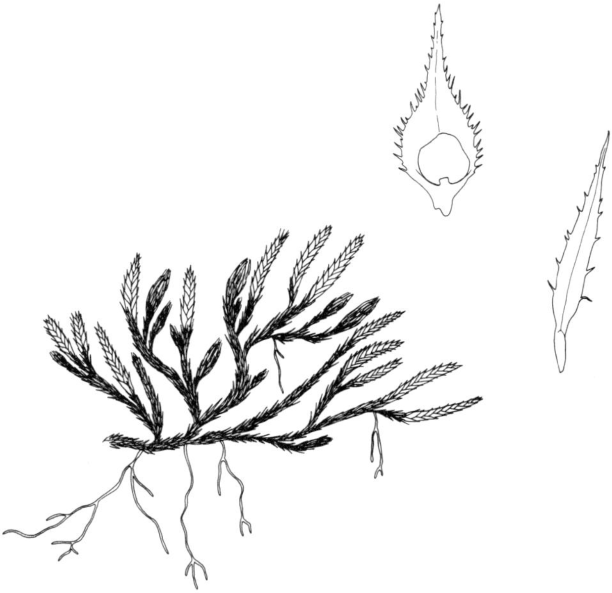

Isoetes

Contributed by W. Carl Taylor

Tufted, grass-like, aquatic, emergent, or terrestrial heterosporous fern-allies. Rootstock corm-like, two- or three-lobed, bearing a cluster of linear, simple leaves above and roots below. Leaves expanded at the base, to bear solitary sporangia. Sporangium walls unpigmented, or brown pigmented in spots or streaks, to completely brown, partly covered by a membranous velum. Megasporangia contain scores of megaspores; microsporangia contain thousands of microspores. Megaspores globose, white, each with an equatorial ridge and three radial ridges converging at the proximal pole, with surfaces echinate, cristate, or nearly smooth. Microspores reniform, grayish in mass, smooth to spinulose.

Isoetes contains about 150 species distributed worldwide. It is a taxonomically difficult genus. Spore morphology revealed by light and scanning electron microscopy (Kott and Britton 1983; Hickey 1986), chromosome numbers (Kott and Britton 1980), electrophoretic profiles of leaf enzymes (Taylor et al. 1985), and spore germination and hybridization experiments (Taylor and Luebke 1986) have provided data useful for improving the taxonomy of *Isoetes*. As with many fern genera, species of *Isoetes* have evolved both by isolation and genetic divergence and by interspecific hybridization and chromosome doubling. It is important to recognize interspecific hybrids and basic diploid and allopolyploid species to make specimens of *Isoetes* easier to identify.

Approximately 24 species of *Isoetes* occur in North America north of Mexico. Species exist as more or less evergreen aquatics in lakes, ponds, and streams; as emergent amphibians along the margins of such bodies of water; or as ephemeral terrestrials, actively growing only when the soil is saturated with water. The base chromosome number in *Isoetes* is 11. The species in New Jersey are diploid ($2n = 22$), tetraploid ($2n = 44$), or decaploid ($2n = 110$).

The Canada goose and mallard duck have been observed eating *Isoetes*. They seem to relish the corm-like rootstock. Waterfowl probably have been important in dispersing plants among lakes and ponds. Aquatic species can be grown in aquaria if temperatures do not exceed 65°F. To remain healthy, plants also require several months of cold dormancy below 40°F. These temperature requirements preclude their use in home fish tanks. Terrestrial species can be grown in shallow pans of soil saturated with water in spring, when the plants are active, allowed to dry out in summer as the plants go dormant, and kept cool and moist in autumn and winter.

Surface ornamentation and size of mature dry megaspores are critical for identification of species. Immature spores are small, fragile, and have an undeveloped spore ornamentation. Plants collected early in the growing season possess immature megaspores and are difficult to identify. Mature megaspores may be found around the rootstock much of the year, but plants collected soon after the growing season are easiest to identify because the spores are well developed and still within sporangia. A hand lens is essential for resolving megaspore ornamentation, and a compound microscope fitted with an ocular micrometer should be used for determining spore size. At least 20 megaspores should be measured to determine average spore size.

Interspecific hybrids can be recognized by their spores of variable size, shape, and surface ornamentation; they may be found where two or more species occur together.

The following key includes six species and one hybrid found in New Jersey. Hybrids can be recognized by variable spore size and identified by comparing their spore ornamentation patterns with the parental species.

Key to Species and More Common Hybrid

1. Megaspores globose, uniform in shape, size, and surface ornamentation (fertile species) **2**

2. Megaspores ornamented with spines .... *I. echinospora*

2. Megaspores ornamented with ridges **3**

3. Megaspores averaging more than 0.60 mm in diameter... *I. lacustris*

3. Megaspores averaging less than 0.60 mm in diameter **4**

4. Megaspores averaging less than 0.50 mm in diameter; ornamented with reticulate ridges or obscure tubercles or ridges **5**

5. Megaspores ornamented with regularly reticulate ridges .... *I. engelmannii*

5. Megaspores ornamented with obscure tubercles or ridges .... *I. melanopoda*

4. Megaspores averaging more than 0.50 mm in diameter; ornamented with isolated or irregularly reticulate crests **6**

6. Megaspores ornamented with high, thin, isolated crests that branch and occasionally anastomose, without densely papillate band along distal side of equatorial ridge .... *I. riparia*

6. Megaspores ornamented with ridges that branch and anastomose, with densely papillate or smooth band along distal side of equatorial ridge .... *I. tuckermanii*

1. Megaspores subglobose, with proximal hemispheres flattened and angular, variable in shape and size, surface ornamentation closely set convoluted ridges .... *I. echinospora* × *engelmannii* (*I.* × *eatonii*)

Isoetes echinospora Durieu

[*I. muricata* Dur.]
[*I. echinospora* var. *braunii* (Dur.) Engelm.]
Spiny-spored Quillwort

Description: Aquatic, occasionally emergent. Leaves subevergreen, up to 25 cm long, bright green, pale toward the base; velum covering <75% of sporangium; sporangium walls occasionally spotted. Megaspores globose, 0.40–0.55 mm in diameter, ornamented with spines. Microspores gray to light brown in mass.

Habitat: Submerged in 5–100 cm of water or partly to completely emergent on wet, sandy, or gravelly shores of slightly acid lakes, ponds, and streams; also tidal mud flats.

Range: Greenland, Labrador to AK, south to NJ, n OH, MI, WI, CO, and CA; Eurasia.

Distribution in New Jersey: Common in northwestern New Jersey, and scattered localities throughout the remainder of the state.

Comments: *I. echinospora* loses most of its leaves in the autumn. Some plants are without visible leaves during the winter, whereas others will bear a few small, dark green overwintering leaves. Typical bright green leaves are produced in the spring as the water warms. Spores mature in late summer, and sporangia soon decay, releasing the spores, which may drift around until the following spring because they appear to require a cold period prior to germination.

I. echinospora hybridizes with at least two other species found in New Jersey. *I.* × *eatonii* (*I. echinospora* × *engelmannii*) is probably the most common hybrid quillwort in New Jersey. *I. echinospora* also hybridizes with *I. riparia* at localities where the two species grow together.

I. echinospora is diploid ($2n = 22$).

Isoetes echinospora Durieu

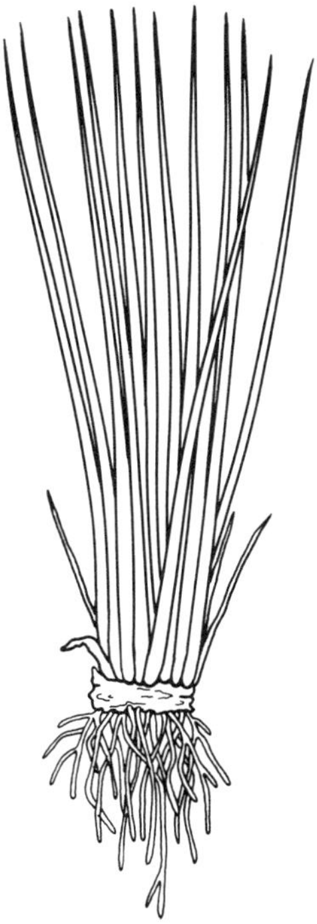

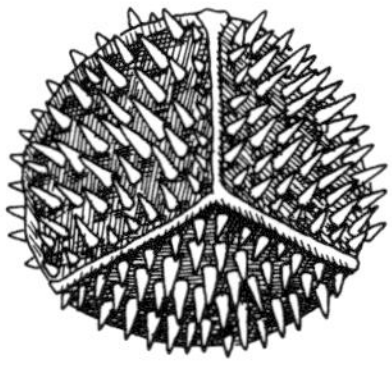

Isoetes lacustris L.

[*I. macrospora* Dur.]
Lake Quillwort

Description: Aquatic. Leaves evergreen, up to 25 cm long, dark green to reddish-green, brownish-white toward base; velum covering <25% of sporangia; sporangium walls occasionally spotted. Megaspores globose, 0.55–0.75 mm in diameter, ornamented with ridges that branch and often anastomose, and have irregular and roughened to rounded and smooth crests, with densely papillate or smooth band along the distal side of the equatorial ridge. Microspores gray in mass.

Habitat: Submerged in 10–900 cm of water on gravelly bottoms of cool, slightly acid lakes and ponds.

Range: Labrador to SK, south to n NJ; disjunct in VA and TN; Europe.

Distribution in New Jersey: A single locality in Sussex County.

Comments: The leaves of *I. lacustris* are normally stiffer and thicker than those of most other *Isoetes*. North American plants of this species have been distinguished as *I. macrospora;* however, both *I. lacustris* and *I. macrospora* are decaploids ($2n = 110$) and have similar leaf and megaspore morphology. In addition electrophoretic studies indicate that both possess a similar genetic constitution. Since they cannot be distinguished except on the basis of geography, it seems prudent to treat them as conspecific.

I. lacustris was discovered in New Jersey by J. L. Edwards in 1937. This species was considered endangered in the state by Fairbrothers and Hough (1973), Snyder and Vivian (1981), and Montgomery (1989), and critically imperiled in the state, although globally secure, by Snyder (1988). In New Jersey *I. lacustris* hybridizes with *I. engelmannii* at the locality for *I. lacustris*.

Isoetes lacustris L.

Isoetes engelmannii A. Braun

Engelmann's Quillwort

Description: Emergent aquatic, occasionally amphibious. Leaves subevergreen, up to 60 cm long, bright green, pale toward base; velum covering <25% of sporangia; sporangia unspotted. Megaspores globose, 0.40–0.50 mm in diameter, ornamented with thin, regularly reticulate ridges. Microspores gray in mass.

Habitat: Submerged, emergent, or occasionally exposed on shores of lakes, ponds, slow-moving streams, and ditches.

Range: NH south to n FL and MS, west to MO.

Distribution in New Jersey: Common in the northern third of the state; scattered elsewhere.

Comments: Spores mature in late summer and they are released as sporangia decay in late summer and fall.

This is the commonest species of quillwort in New Jersey. It occurs in many small ponds and slow-moving streams; it is the only species likely to be found in farm ponds. *I. engelmannii* seems somewhat tolerant of pollution, as it grows in streams in industrial areas and golf courses in central New Jersey.

I. engelmannii hybridizes with *I. lacustris* and *I. riparia* at localities where these species grow together in New Jersey.

The specific epithet commemorates G. Engelmann, who first collected this species near St. Louis, MO, in 1842.

I. engelmannii is diploid ($2n = 22$) (Wagner and Wagner 1966).

Isoetes engelmannii A. Braun

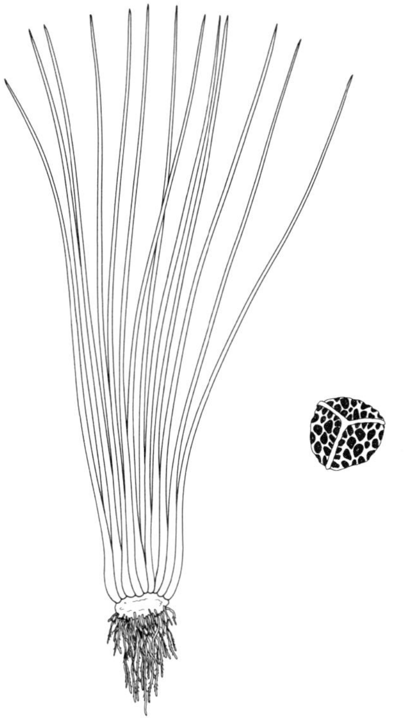

Isoetes melanopoda Gay & Durieu

Black-footed Quillwort

Description: Emergent to terrestrial. Leaves deciduous, up to 40 cm long, bright green, pale to lustrous black toward the base; velum covering <75% of sporangia; sporangium walls spotted. Megaspores globose, 0.28–0.44 mm in diameter, ornamented with obscure tubercles or ridges. Microspores gray in mass.

Habitat: In clay-bottomed depressions that are flooded in spring but normally dry by early summer, meadows and fields, ditches, and thin soil pockets on sandstone and granite outcrops.

Range: VA to SD, south to AR, TX, and UT, east to GA; widespread in the Midwest; disjunct in s NJ.

Distribution in New Jersey: Two "vernal ponds" in Cape May County.

Comments: The New Jersey populations are isolated (disjunct) several hundred miles northeast of the main range of this species. *I. melanopoda* was considered endangered in New Jersey by Fairbrothers and Hough (1973), Snyder and Vivian (1981), and Montgomery (1989), and critically imperiled in the state, although globally secure, by the New Jersey Natural Heritage Program (Snyder 1988). The ponds where the quillworts occur contain several additional species that are rare or endangered in New Jersey. The habitat is unique in the state: acid depressions underlain by impermeable clay, in the otherwise very permeable sand of the Outer Coastal Plain (Montgomery and Fairbrothers 1963). Spores mature in late spring as the soil dries and plants become dormant. By midsummer leaves have withered and plants are no longer apparent.

I. melanopoda is diploid ($2n = 22$) (W. C. Taylor unpublished data).

Isoetes melanopoda Gay & Durieu

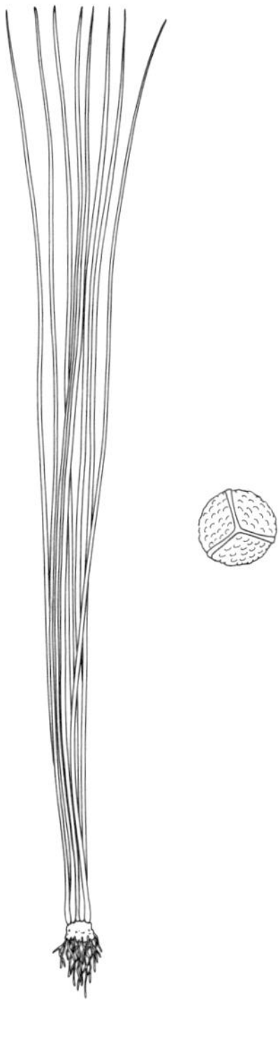

Isoetes riparia Engelmann

[*I. riparia* var. *canadensis* Engelmann]
Shore Quillwort

Description: Aquatic, occasionally emergent. Leaves evergreen, up to 35 cm long, bright green, pale toward base; velum covering <50% of sporangia; sporangium walls spotted to completely brown. Megaspores globose, 0.45–0.65 mm in diameter, ornamented with high, thin, isolated crests that branch and occasionally anastomose, without densely papillate or smooth band along the distal side of the equatorial ridge. Microspores gray to light brown in mass.

Habitat: Submerged on sandy to gravelly bottoms of lakes, ponds, and tidal shores of rivers where plants are exposed at low tide.

Range: s ME to s ON, south along the eastern seaboard to n FL.

Distribution in New Jersey: Scattered localities in the northern counties; along the tidal Delaware, Mullica, and Tuckahoe Rivers in southern New Jersey.

Comments: Spores of this species mature in late summer and are released as the sporangia decay in late summer and autumn.

I. riparia was considered rare in the state and other parts of its range by the New Jersey Natural Heritage Program (Snyder 1988). Pollution and wetland disturbance have eliminated this and many other wetland species from parts of New Jersey, especially along tidal rivers.

I. riparia hybridizes with *I. echinospora* in localities where the two grow together. Evidence suggests that it is an allotetraploid, formed by chromosome doubling of diploid species *I. echinospora* × *I. engelmannii* (= *I.* × *eatonii*) (Taylor et al. 1985).

I. riparia is tetraploid ($2n = 44$).

Isoetes riparia Engelmann

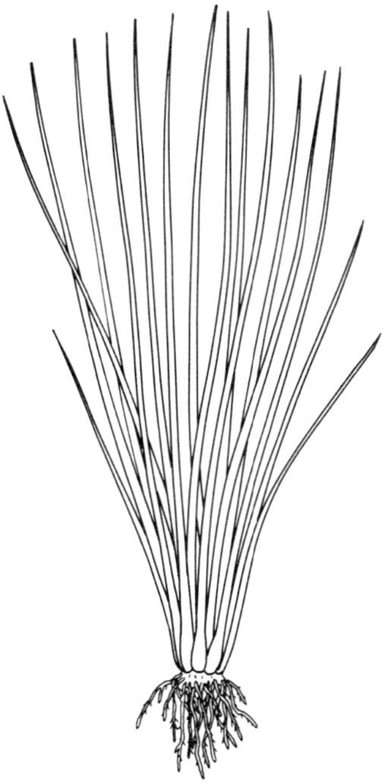

Isoetes tuckermanii A. Braun

Tuckerman's Quillwort

Description: Aquatic. Leaves evergreen, up to 20 cm long, olive to grayish-olive or reddish-brown, pale toward the base; velum covering <50% of sporangia; sporangium walls often spotted. Megaspores globose, 0.45–0.65 mm in diameter, ornamented with ridges that branch and anastomose, with densely papillate or smooth band along distal side of equatorial ridge. Microspores gray in mass.

Habitat: Lakes, ponds, and river shores.

Range: NF west to PQ, south to NJ.

Distribution in New Jersey: A single locality in Warren County.

Comments: *I. tuckermanii* has gone unrecognized from New Jersey, although material was collected in 1947. New Jersey is the southern limit of geographic distribution for this species. It is common in ponds in New England and the Maritime Provinces of Canada.

Spores of this species mature in late summer and are released in late summer and autumn as the sporangia decay.

Plants with megaspores bearing low, bold, rounded ridges and a smooth girdling band along the equatorial ridge have been called *I. acadiensis* (Kott 1981), but mass collections have shown that within a single population megaspore ornamentation grades from typical *I. tuckermanii* to *I. acadiensis*.

It hybridizes with several species in New England, but no such hybrid has been reported from New Jersey.

I. tuckermanii is a tetraploid species ($2n = 44$).

Isoetes tuckermanii A. Braun

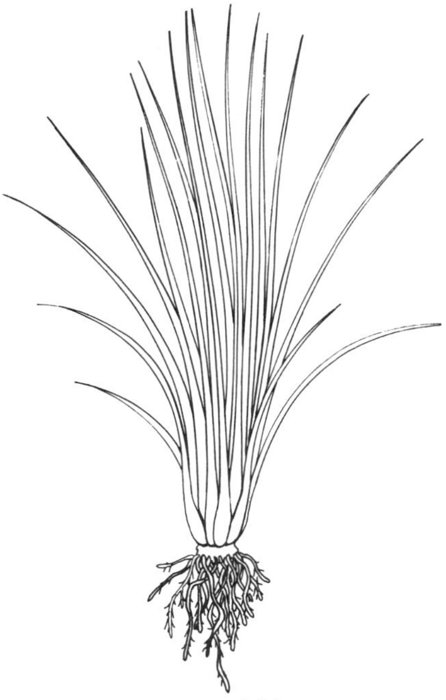

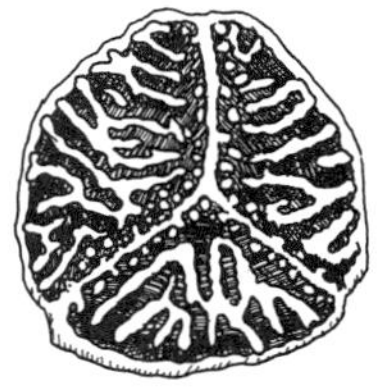

Isoetes × eatonii Dodge
I. echinospora × engelmannii

Eaton's Quillwort

Description: Aquatic, occasionally emergent. Leaves evergreen to subevergreen, up to 50 cm long, bright green, pale toward base; velum covering <50% of sporangia; sporangium walls occasionally spotted. Megaspores subglobose with proximal hemispheres flattened and angular, variable in size and shape, mostly 0.30–0.55 mm in diameter, ornamented with closely set, convoluted ridges. Microspores gray in mass.

Habitat: Submerged in 10–100 cm of water in lakes.

Range: Southern NH to s ON, south to n NJ.

Distribution in New Jersey: Several localities from Sussex. Morris, and Passaic counties.

Comments: Evidence from spore morphology, spore germination, and hybridization experiments and from electrophoretic profiles indicates that *I. × eatonii* is a sterile interspecific hybrid between *I. echinospora* and *I. engelmannii.* Unlike the uniform globose megaspores of fertile species, the variable subglobose megaspores of *I.* × eatonii do not germinate in culture. The spores also appear in wet mounts to be devoid of protoplasm. Crosses between *I. echinospora* and *I. engelmannii* are easily made in culture, and these artificial hybrids can be compared with the natural plants as to morphology and chemistry. In addition, *I. × eatonii* occurs within the partly sympatric ranges of *I. echinospora* and *I. engelmannii* (Taylor et al. 1985).

This taxon, considered as a species, was listed as endangered in New Jersey by Fairbrothers and Hough (1973) and Snyder and Vivian (1981). However, hybrids are not usually considered for listing if they are sterile and can be formed by crosses of the parents. Montgomery (1989) pointed out that this is the case with *I. × eatonii.*

Isoetes × *eatonii* Dodge

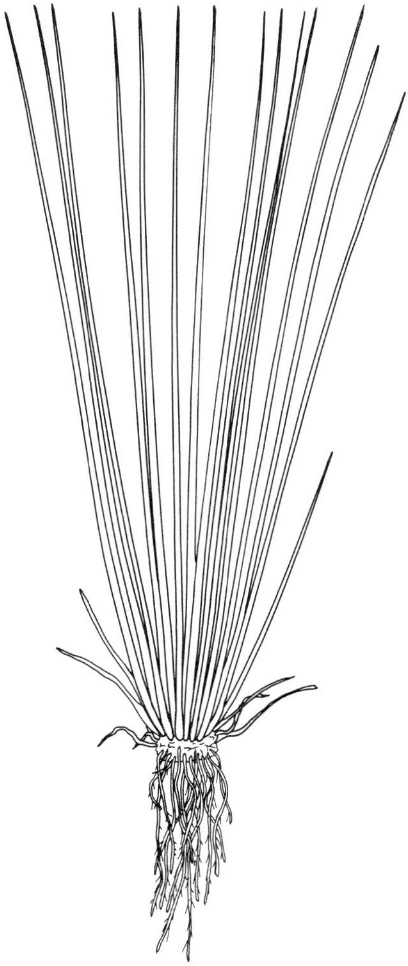

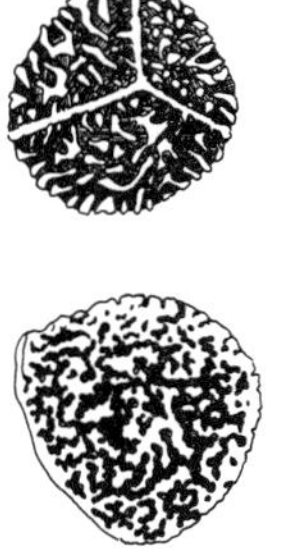

Additional Isoetes Hybrids

Three additional hybrids of *Isoetes* have been found from New Jersey, but all are known from fewer than five stations. All of these hybrids are recognized by their megaspores, which are variable in shape and size. Sporangia contain many subglobose megaspores and a smaller number of globose, hemispherical, lunate, or dumbbell-shaped forms. Megaspore ornamentation patterns of the hybrids appear somewhat intermediate between the parents. In addition, microspores are seen to be mostly empty when viewed in wet mounts under the compound microscope. The following occur in New Jersey:

Isoetes × *brittonii* Brunton & Taylor
I. engelmannii × *riparia*

This hybrid is known from one locality in Sussex County and two localities in Morris County. It has been collected only once at each station.

Isoetes engelmannii × *lacustris*

This hybrid is known from a single locality in Sussex County, which is also the only known station for *I. lacustris* in the state.

Isoetes × *dodgei* Eaton
I. echinospora × *riparia*

This hybrid has been collected once along the Delaware River, in Camden County.

Equisetum

Small-to-large plants with creeping rhizomes; erect stems ribbed, hollow, and jointed; leaves whorled, fused into a sheath around the stem above the joint (node), and projecting as teeth on the sheath; branches alternating with the leaves, or stems unbranched; sporangia borne on specialized sporangiophores in strobili.

The genus *Equisetum* is morphologically distinctive and has existed for at least 200 million years. Fossil relatives were among the Paleozoic coal-forming plants, and they grew to tree size. There are 15 living species, 11 in North America, and 6 in New Jersey; hybrids exist, and 3 occur in New Jersey.

There are two groups, usually treated as subgenera. Excellent recent monographs of each group have been published by R. L. Hauke (1963, 1978). Our treatment is based on these monographs.

The chromosome number for all species examined is $n = 108$ (Hauke 1978).

Key to Species and Hybrids

1. Stems (above ground) annual, usually with whorls of branches; cones rounded at apex: subgenus *Equisetum* **2**

2. Stems regularly branched; first branch internode mostly longer than main stem sheath **3**

3. Stem sheath teeth red, lacking white margins; branches branched .... *E. sylvaticum*

3. Stem sheath teeth brown with white margins; branches unbranched **4**

4. Branch sheath teeth deltoid; main stem teeth with broad white margins; branches drooping to spreading.......... *E. pratense*

4. Branch sheath teeth narrow-triangular; main stem teeth with narrow white margins; branches usually ascending................ .... *E. arvense*

2. Stems unbranched, irregularly branched, or with regular short branches; first branch internode shorter than to barely equaling main stem sheath.. **5**

5. Ridges on main stem 12–24; sheaths tightly appressed to stem; spores normal .... *E. fluviatile*

5. Ridges on main stem 10–14; sheaths flaring outward from main stem; spores abortive .... *E. × litorale*

1. Stems perennial, evergreen, usually not branched (damaged plants may branch); cones pointed at apex: subgenus *Hippochaete*.............. **6**

6. Teeth of stem sheath mostly falling off; stems 3–17 mm in diameter **7**

7. Upper stem sheaths white, with black band near the middle; spores normal .... *E. hyemale* var. *affine*
7. Upper stem sheaths green, lacking a dark band; spores abortive .... *E.* × *ferrissii*

6. Teeth of stem sheath persistent; stems 0.5–5.0 mm in diameter...... **8**
8. Sheaths green with a black rim; teeth obtuse, 1–2 mm long, with deciduous filiform tip; spores normal .... *E. variegatum*
8. Sheaths with black band and white tip or black above; teeth acute, up to 3 mm long with a persistent filiform tip; spores abortive .... *E.* × *trachyodon*

Equisetum sylvaticum L.

Woodland Horsetail

Description: Stems annual, branched in regular whorls with the branches again branched; 25–70 cm tall, 1.5–3.0 mm in diameter, with 8–18 ridges; first internode longer than stem sheath; sheath reddish brown, papery, with brown teeth usually cohering in three or four groups; fertile stems at first nongreen, becoming green and branched after spores are shed; cones rounded at apex.

Habitat: Swamps or moist upland woods, and seepy wooded slopes.

Range: Greenland and Labrador to AK, south to NJ, VA, WV, OH, IA, NE, and WA; also Eurasia.

Distribution in New Jersey: Scattered in the northern counties (no recent records for Bergen or Essex); not found south of the Wisconsin glacial boundary.

Comments: *E. sylvaticum* is the most attractive species of horsetail, but it is difficult to transplant and establish. It is the only species in the genus with branches rebranching.

Equisetum sylvaticum L.

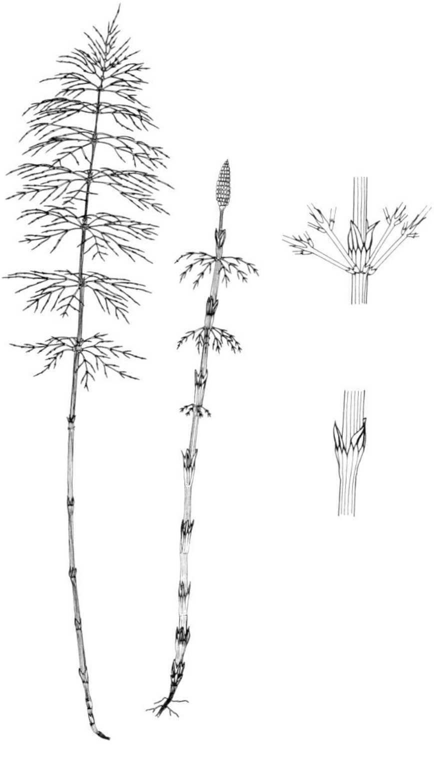

Equisetum pratense Ehrh.

Meadow Horsetail

Description: Stems annual, regularly branched with spreading to drooping branches, 16–53 cm tall, 1–3 mm in diameter, with 8–18 ridges; first internode longer than the sheath (at least in the upper part of the stem); sheath 3–5 mm long, teeth narrow, with white border as wide or wider than the dark center; fertile stems at first non-green and unbranched, becoming green and branched after spores are shed; cones rounded at apex.

Habitat: Stream banks and springy moist woods.

Range: NF to AK, south to n NJ, MI, IA, ND, and BC; also Eurasia.

Distribution in New Jersey: Rare in the state, where it reaches the southern limit of its geographic range.

Comments: This species was considered endangered in New Jersey by Fairbrothers and Hough (1973), Snyder and Vivian (1981), and Montgomery (1989). It was considered critically imperiled in the state, although globally secure, by the New Jersey Natural Heritage Program (Snyder 1988). The stations in Bergen (1865) and Passaic (1944) counties have disappeared, and only two proximate stations in Morris County remain. This is another example of a species that is rare because it is at the limit of its geographic distribution in New Jersey and requires a specialized habitat in the state.

Equisetum pratense Ehrh.

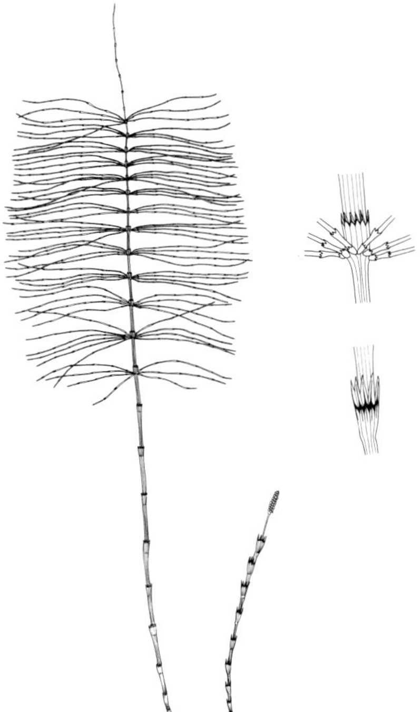

Equisetum arvense L.

Field Horsetail

Description: Stems annual, the vegetative ones regularly branched with ascending to horizontal branches, 10–100 cm tall, 0.8–4.5 mm in diameter, with 4–14 ridges; first branch internode longer than subtending sheath; sheaths green, teeth narrow with white margins narrower than the dark central stripe; fertile stems in early spring, withering soon after shedding spores, unbranched, succulent, usually not becoming green; cones rounded at apex.

Habitat: Moist open habitats, especially in disturbed sandy soil such as roadsides, railroad banks, stream banks, open fields (including cultivated fields), and open woods.

Range: Arctic Canada, south to SC, and west to CA; also Eurasia.

Distribution in New Jersey: Abundant; recorded in all 21 counties; recent records in all except Cumberland County.

Comments: *E. arvense* is an extremely variable plant (like many species of disturbed habitats). There are graceful upright shade forms, tall forms branched only near the top, and bushy forms with upright branches. Variations within the fertile and vegetative stems also occur. Hauke (1966) studied the diversity and concluded that none of the morphological variations were correlated with each other nor with geographic distribution; thus no subspecific taxa should be recognized.

Equisetum arvense L.

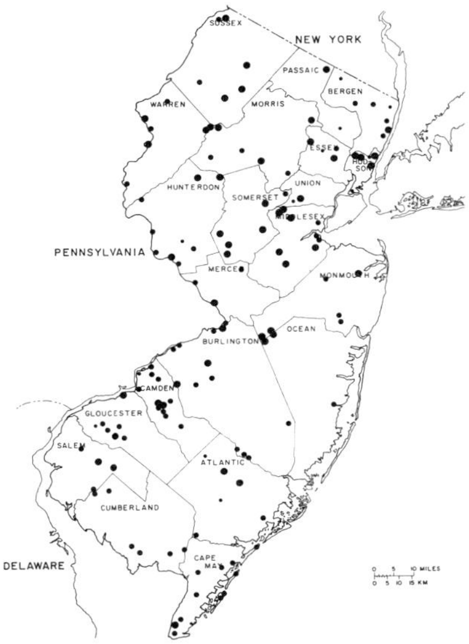

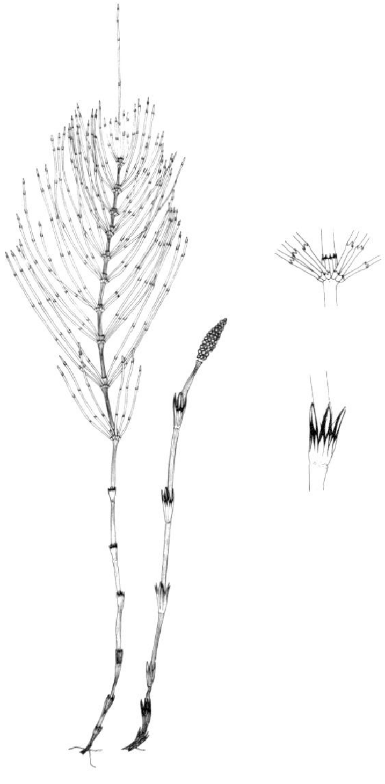

Equisetum fluviatile L.

Water Horsetail

Description: Stems annual, unbranched, irregularly branched, or with whorls of short branches from the middle nodes; 35–115 cm tall, 2.5–9.0 mm in diameter, with 12–24 ridges; first branch internode shorter than main stem sheath; sheath appressed to stem, with narrow dark teeth; fertile stems like the vegetative; cones rounded at apex.

Habitat: Shallow water of marsh borders, wet meadows, edges of lakes and rivers.

Range: NF to AK, south to DE, VA, OH, IA, NE, WY, and OR; also Eurasia.

Distribution in New Jersey: Scattered in the northern counties; extends south along the Delaware River (mostly older records).

Comments: The regularly branched and the unbranched forms look very different and were named as separate species by Linnaeus (Hauke 1978). Since the morphological characters intergrade, nearly all recent treatments have considered them to be a single species. There is a tendency for the unbranched stems to be produced early in the spring and for later stems to bear regular whorls of branches.

This is an attractive species and, if contained, makes an interesting display around the margin of a pond.

Equisetum fluviatile L.

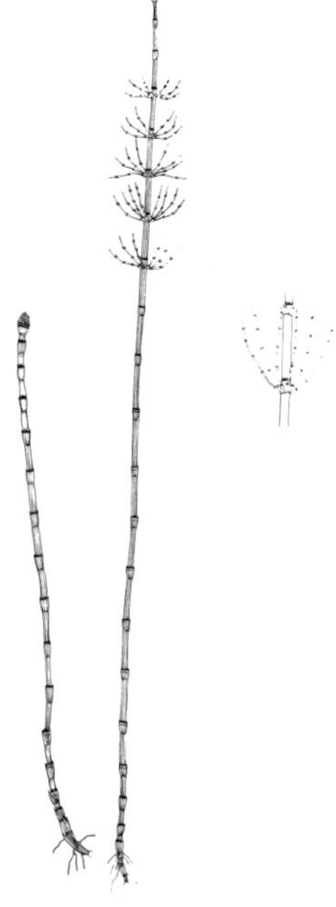

Equisetum × *litorale* Kuehl.
E. arvense × *fluviatile*

Shore Horsetail

Description: Stems annual, unbranched or irregularly branched, 20–100 cm tall, 1.2–4.0 mm in diameter, with 10–14 ridges; first branch internode (if present) shorter than or nearly equaling stem sheath; sheaths somewhat flaring at top, with black teeth with very narrow white margins; stems all alike, but fertile ones rare; spores abortive.

Habitat: Along river and lake shores, alluvial woods.

Range: NF to NJ, west to BC and WA; also Europe.

Distribution in New Jersey: Several stations on limestone ledges along the Delaware and Pequest rivers.

Comments: Peck (1980) studied the population dynamics of stands of this hybrid in the upper Midwest. He reported that stands, probably originating as a single sterile hybrid plant, could spread by asexual reproduction for at least 80 years. In New Jersey populations along the Delaware River have been known since 1884.

Equisetum palustre was recorded from New Jersey by Chrysler and Edwards based on a specimen collected by W. S. Disbrow and presently located at the Newark Museum. Examination of this specimen by J. D. Montgomery and of a photocopy by R. L. Hauke indicates that the specimen is *E.* × *litorale* rather than *E. palustre,* which is a more northern species.

Equisetum × *litorale* Keuhl.

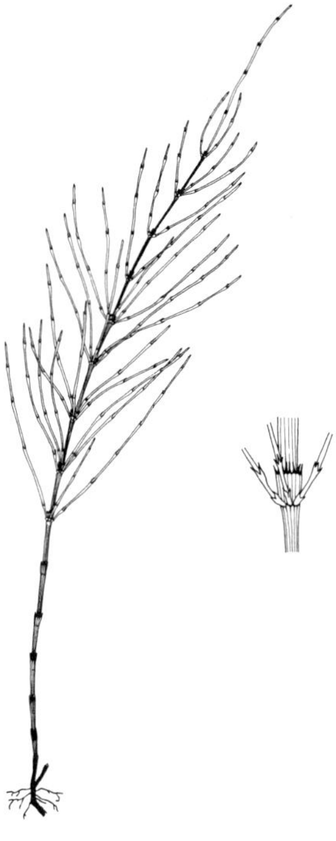

Equisetum hyemale L.

E. hyemale var. *affine* (Engelm.) A. A. Eaton
[*E. praealtum* Raf.]
Scouring-rush

Description: Stems evergreen, normally unbranched, 20–200 cm tall, 3.0–17.5 mm in diameter, with 14–50 ridges; sheath with a dark band across the middle and tan to ashy-gray above, appressed to stem; sheath teeth deciduous; stems all alike, cones apiculate.

Habitat: Sandy or gravelly damp slopes, including railroad banks and roadsides, wooded borders, and stream banks.

Range: Labrador to AK, throughout the United States and Mexico to Guatemala; var. *hyemale* occurs in Eurasia.

Distribution in New Jersey: Common in the northern counties, and scattered on the Piedmont and Inner Coastal Plain in southern New Jersey.

Comments: The name scouring-rush comes from the use of the rough stems containing silica to scour pots. This species is fairly easily transplanted and can be used for erosion control on embankments if they are not too dry. The stems are striking when seen in a dense stand. Like other members of the genus, *E. hyemale* can become invasive; therefore, it should not be used in landscaping where it can not be contained.

Equisetum hyemale L.

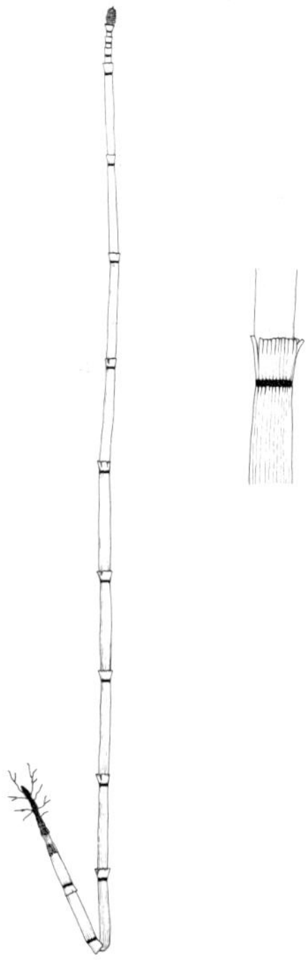

Equisetum × *ferrissii* Clute
E. hyemale var. *affine* × *laevigatum*

Intermediate Scouring-rush

Description: Stems evergreen, unbranched, 25–180 cm tall, 3–11 mm in diameter with 14–32 ridges; sheaths on lower part of the stem with dark band and white above (like *E. hyemale*) and on upper part of stem all green (like *E. variegatum*), 7–17 mm long; sheath teeth deciduous or partly persistent, leaving a dark rim on the tip of the sheath; stems all alike; cones apiculate, spores abortive.

Habitat: Gravelly or sandy calcareous shores, railroad embankments, roadsides, and fields.

Range: CT, NY, and NJ, to VA, west to BC, CA, and Baja California, Mexico.

Distribution in New Jersey: This *Equisetum* hybrid has been known from New Jersey since 1886, and one extant colony has been known since 1907.

Comments: *E. ferrissii* has been confused with *E. laevigatum* (one of the parents) (Chrysler and Edwards 1947), but this species has annual stems, green unbanded sheaths throughout, and rounded cones. *E.* × *ferrissii* occurs well to the east of the range of *E. laevigatum* in the United States. Hauke (1963) concluded that *E.* × *ferrissii* has spread by rhizomes transported along watercourses, canals, and railroads from the Great Lakes to New York, New Jersey, and south to Virginia.

Equisetum × *ferrissii* Clute

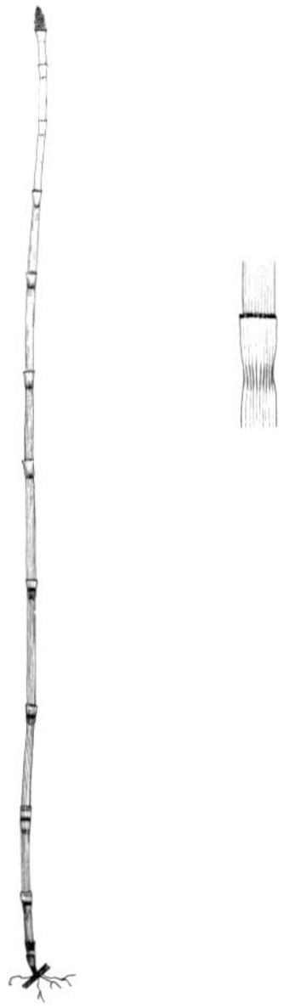

Equisetum variegatum Schleich.

Variegated Scouring-rush

Description: Stems evergreen, unbranched, 6–45 cm tall, 0.5–3.0 mm in diameter, with 3–12 ridges; sheaths green with black rim, somewhat flaring; sheath teeth persistent, obtuse, with broad white margins and dark central stripe, narrow tip deciduous; stems all alike; cones apiculate.

Habitat: In damp soil of river banks, ditches, and meadows; often in calcareous soil.

Range: Greenland, Labrador, and AK, south to n NJ, nw PA, MI, MN, CO, and CA; also Eurasia.

Distribution in New Jersey: Rare, now known from three stations in Warren and Morris counties.

Comments: *E. variegatum* was discovered by V. Abraitys in Warren County in 1977 (Montgomery 1981). A second colony was found by J. D. Montgomery in 1983. Both areas had been visited by botanists not many years before the discoveries; therefore, this species is probably a recent introduction into New Jersey. There was an earlier report of this species by C. F. Austin from "Closter, Bergen Co."; both Chrysler and Edwards (1947) and Montgomery (1981) concluded that this was an error or possibly a specimen from a cultivated plant.

E. variegatum was considered endangered in New Jersey by Snyder and Vivian (1981) and Montgomery (1989); it was considered critically imperiled in the state, although globally secure, by the New Jersey Natural Heritage Program (Snyder 1988).

Equisetum variegatum Schleich.

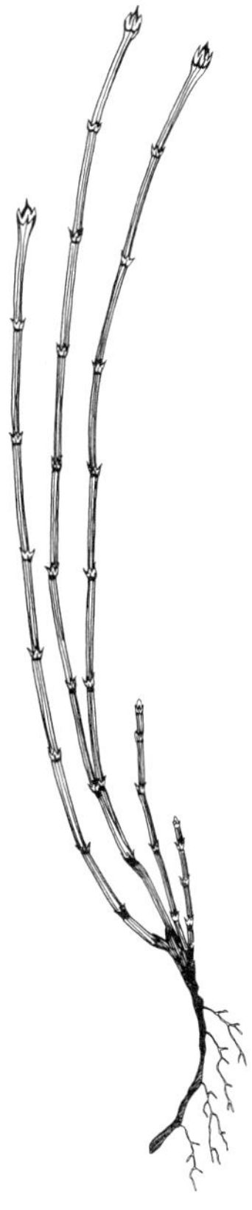

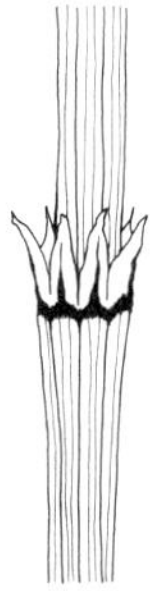

Equisetum × *trachyodon* A. Br.
E. hyemale var. *affine* × *E. variegatum*

[*E. variegatum* var. *jessupi* A. A. Eaton]
Hybrid Scouring-rush

Description: Stems evergreen, unbranched, 20–85 cm tall, 1.5–5.0 mm in diameter, with 7–16 ridges; sheath appressed, the upper part black or with a black band and white tip; sheath teeth persistent, acute, up to 3 mm long, with a brown center and white margins; stems all alike, cones apiculate, spores abortive.

Habitat: Edges of rivers, lakes, and swamps.

Range: NF to BC, south to NY, NJ, OH, IN, WY, MT, and OR; also Iceland and Europe.

Distribution in New Jersey: Crevices of limestone or shale along the Delaware River, Sussex County.

Comments: *E.* × *trachyodon* was first collected in New Jersey by J. L. Edwards in 1950. Montgomery (1981) discussed the appearance of this hybrid (and *E. variegatum*) in New Jersey and concluded that this hybrid is a recent introduction to the state. The Delaware River has its source in the Catskill Mountains of New York, from where this hybrid may have spread downriver by vegetative material. *E. variegatum* occurs in the Catskills, but it appears that this hybrid occurred in New Jersey before the species did.

Equisetum × *trachyodon* A. Br.

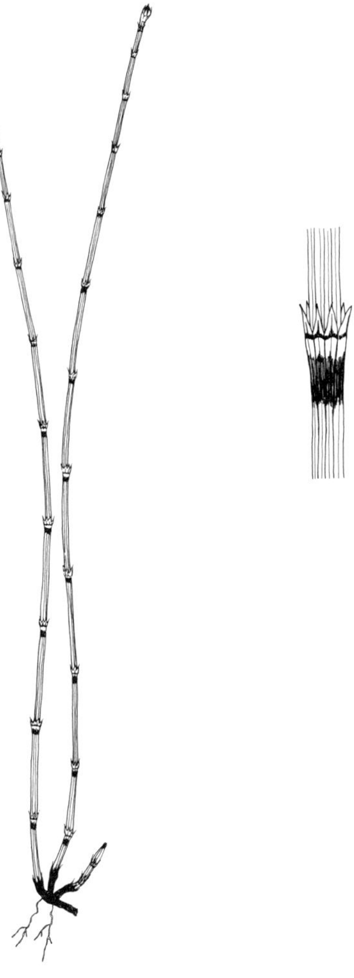

Ferns

Botrychium

Small to medium-sized ferns. Rhizomes very short, erect, with fleshy roots; the plant producing a single frond each year (a second frond may be produced if the first is damaged). Fronds with basal fertile and apical vegetative segments; the vegetative portion barely lobed to quadripinnate. Sporangia borne in a branched panicle, large and globose, opening by a slit, and with many spores.

Botrychium contains three subgenera (Clausen 1938), all represented in New Jersey. They differ in morphology, phenology, and ecology. The grape ferns, subgenus *Sceptridium,* are medium-sized plants, with bi- to tripinnate fronds produced in midsummer, and remaining green or turning bronze-colored through the following spring; they shed spores in early autumn. The moonworts, subgenus *Botrychium,* are small plants, with pinnatifid to bipinnate fronds, produced in late spring (in New Jersey) and surviving not much longer than when the spores are shed, in midsummer. The rattlesnake ferns, subgenus *Osmundopteris,* are represented in New Jersey by a single species, with fairly large fronds, at least tripinnate, produced in spring and lasting into late summer; spores are shed in early summer. Most grape ferns and moonworts grow in abandoned fields and second growth woods; they are not usually found in mature woods, except *B. oneidense*. The rattlesnake fern is a plant of young-to-mature woodlands. *Botrychium* species often grow together (Wagner and Wagner 1983), and in New Jersey it is not uncommon to find four or five species in an abandoned field or open woods.

Except for *B. virginianum,* which can be grown in woodland gardens, grape ferns and moonworts are very difficult to cultivate. This is presumably due to the mycorrhizae, or associated fungi, growing in the roots and probably aiding in nutrition of the plant.

Key to Species

1. Fronds herbaceous in texture, deciduous, arising in spring and shedding spores in early summer; fertile segments arising near base of vegetative blades, well above soil level except in very small plants **2**

2. Vegetative blades tri- or quadripinnate, 20–50 cm long .... *B. virginianum*

2. Vegetative blades simple to bipinnate, 2–20 cm long **3**

3. Vegetative blades triangular, sessile; segments acute .... *B. lanceolatum*

3. Vegetative blades elliptic or ovate, short-stalked; segments obtuse or rounded **4**

4. Vegetative blades pinnate to bipinnate, segments lanceolate to ovate, with a midrib .... *B. matricariifolium*

4. Vegetative blades almost entire to pinnatifid, segments spoon-shaped to fan-shaped, without a midrib .... *B. simplex*

1. Fronds leathery in texture, wintergreen, arising in mid- to late summer, shedding spores in late summer; fertile segments arising near base of stipe (at or below soil level) **5**

5. Terminal segments rounded at apex; margins with rounded teeth (crenate).. **6**

6. Blade segments all about alike in shape, terminal segments not elongate; plants of fields and open woods.............. *B. multifidum*

6. Blade with terminal segments elongate; plants of alluvial woods .... *B. oneidense*

5. Terminal segments acute at apex; margins with pointed teeth (serrate) .... *B. dissectum*

Botrychium virginianum **(L.) Sw.**

Rattlesnake Fern

Description: Fronds produced in spring, withering in autumn; spores shed in late spring; vegetative blades tri- or quadripinnate, 15–50 cm long, triangular, sessile; fertile segments, if present, overtopping the vegetative blades.

Habitat: Rich loamy woods and moist wooded slopes, usually under northern hardwoods or hemlock-hardwoods.

Range: PE and PQ to s BC, south to s Mexico; also e Asia; a variety occurs in the w United States and Europe.

Distribution in New Jersey: Common in the Ridge and Valley, Highlands, and Inner Coastal Plain; scattered on the Piedmont; rare in the Pinelands and Outer Coastal Plain.

Comments: Chrysler and Edwards (1947) noted that this species is one of several plants that has a common name connected, for no apparent reason, with snakes. Their remark "this is too beautiful a plant to be so maligned" is an interesting comment on changes in the attitude toward New Jersey's wildlife in the last 45 years or so. The name rattlesnake fern is supposed to originate from a resemblance of the fertile segment to the end of the tail of a rattlesnake, but this is much more appropriate for *Ophioglossum* than *Botrychium*.

The chromosome number of *B. virginianum* is $2n = 184$. This is different from other numbers of the genus and reflects the differences in the subgenera of *Botrychium*.

Botrychium virginianum (L.) Sw.

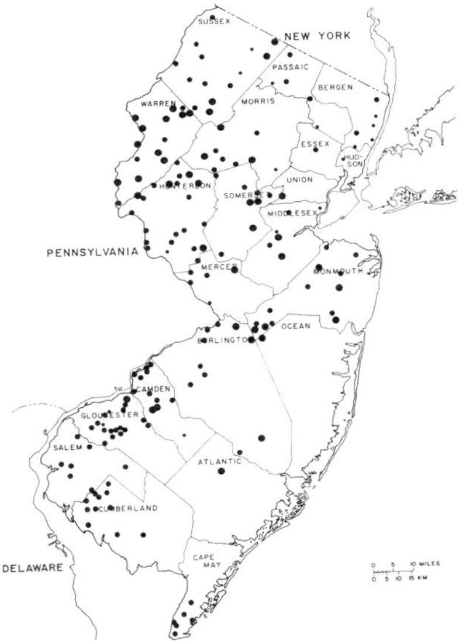

Botrychium lanceolatum (Gmel.) Angstr. subsp. *angustisegmentum* (Paese & Moore) Clausen

Lance-leaved Moonwort

Description: Fronds produced in late spring, usually withering soon after spores are shed in early summer; vegetative blades pinnate-pinnatifid to bipinnate-pinnatifid, 1–6 cm long, triangular, bright shiny green, sessile; ultimate segments lanceolate with acute apices, midrib present; fertile segments (usually present) slightly longer than the vegetative blades.

Habitat: Moist to wet woods; sandy to loamy woods in southern New Jersey.

Range: NF to BC and s AK, south to NJ, PA, NC, and OH; also in Europe.

Distribution in New Jersey: Scattered over the northwestern counties; rare (only older records) in northeastern and southwestern counties.

Comments: In the northeastern United States, this species is represented by subsp. *angustisegmentum;* the subsp. *lanceolatum* is more northern and western.

The chromosome number of *B. lanceolatum* is $2n = 90$.

Botrychium lanceolatum (Gmel.) Angstr.

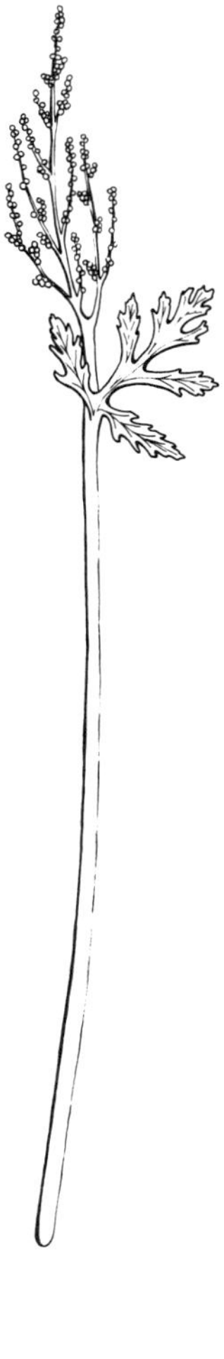

Botrychium matricariifolium A. Br.

Daisy-leaved Moonwort

Description: Fronds produced in late spring, withering soon after spores are shed in early summer; vegetative blades 1–10 cm long, pinnate-pinnatifid to bipinnate-pinnatifid, lanceolate to ovate, light, dull green, short-stalked; ultimate segments ovate, with obtuse apices, midrib present; fertile segments (usually present) overtopping the vegetative segments.

Habitat: Fairly dry to moist open woods, thickets, and abandoned fields; also in conifer plantations.

Range: NF to BC, south to Long Island, NY, NJ, VA, NC, TN, WI, SD, and ID; also in Eurasia.

Distribution in New Jersey: Common in the northern counties (but many records older); scattered over the southwestern counties; rare in the Pinelands; absent in southernmost New Jersey.

Comments: This is the commonest moonwort in New Jersey, but plants are frequently overlooked in leaf litter and dense shrubs. Like other *Botrychium* species, this is a successional species, much like a goldenrod, and can be found in the perennial and shrub stages of secondary succession, as well as in young forest. The plants often associate with and look like wood sorrel (*Rumex acetosella*). Plants are also frequently found under apple or cherry trees in abandoned orchards. An interesting account of the distribution of this species in New Jersey was presented by B. Long (1922).

B. matricariifolium is tetraploid, with $2n = 180$.

Botrychium matricariifolium A. Br.

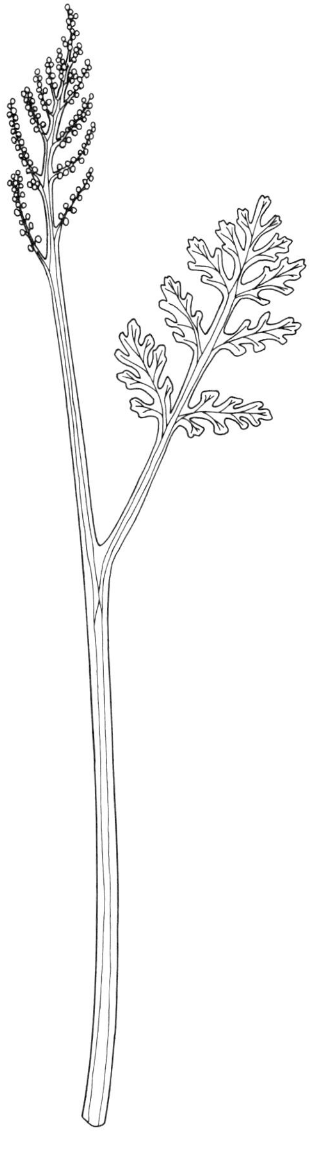

Botrychium simplex E. Hitchc.

Least Moonwort

Description: Fronds produced in late spring, withering soon after spores are shed in early summer; vegetative blades nearly simple, lobed, pinnatifid or pinnate, 1–6 cm long, linear to ovate-lanceolate, dull light to medium green, short- to long-stalked; segments, if any, fan-shaped, blunt, lacking a midrib; fertile segments (usually present) shorter to longer than vegetative blades.

Habitat: Open damp fields or seepage to dry or moist woods and wooded slopes at borders of swamps, in leaf litter or moss mats.

Range: NF to s BC, south to n NJ, n PA, VA, NC, IN, NM, and CA; also Eurasia.

Distribution in New Jersey: Sussex, Warren, Morris, and Passaic counties; rare.

Comments: Three varieties of *B. simplex* have been referred to New Jersey's flora. These are perhaps better considered as ecological forms, as suggested by Wagner (1963). The taxon growing in open, exposed habitats is var. *simplex,* which has leathery blades with fan-shaped segments. The taxon growing in damp woods and edges of swamps is var. *tenebrosum* (A. A. Eaton) Clausen, and is very slender, with elongate simple to barely lobed vegetative blades, with a long stalk above the junction with the fertile segment. The taxon growing in drier woods is var. *laxifolium* Clausen, which has more compact, deeply lobed blades, with a short stalk. All of these taxa are rare in New Jersey, which is near the southern geographic limit of this species (Montgomery 1982b, 1989). The first record for the type variant in New Jersey was in 1981, collected and reported by Snyder (1985).

B. simplex is diploid, with $2n = 90$.

Botrychium simplex E. Hitchc.

Botrychium dissectum Spreng.

[*B. obliquum* Muhl.]
Cut-leaved Grape Fern

Description: Fronds produced in midsummer, remaining green, or turning bronze in winter, and lasting to the following summer; spores shed in late summer; vegetative blades at least tripinnate, triangular; segments sharply and irregularly serrate, with acute apices; terminal segments longer than broad, and much more elongate than lateral segments of the pinnae.

Habitat: Moist woods, thickets, and abandoned fields.

Range: NS west to MN, south to FL, LA, and TX.

Distribution in New Jersey: Common throughout the state; records from all 21 counties.

Comments: Two forms of *B. dissectum* occur in New Jersey, usually in the same population: *B. dissectum* forma *obliquum* (Muhl.) Clute is the more common form, with normally expanded blades; forma *dissectum,* in the extreme form, has very skeletonized blades, with all segments cut into linear teeth. The two forms intergrade, and many intermediates can be found. In large populations a few plants of f. *dissectum* are scattered among numerous plants of f. *obliquum,* with occasional individuals of intermediate morphology. All individuals produce good spores, indicating that this variation is not the result of hybridization but of diversity in morphology.

The chromosome number of *B. dissectum* is $2n = 180$ (tetraploid).

B. biternatum (Sav.) Underw. was attributed to Cape May County by O. H. Brown and E. T. Wherry (1970). This species is similar to *B. dissectum* but of more southern geographic distribution. No herbarium specimen has been found to substantiate the presence of *B. biternatum* in New Jersey.

Botrychium dissectum Spreng.

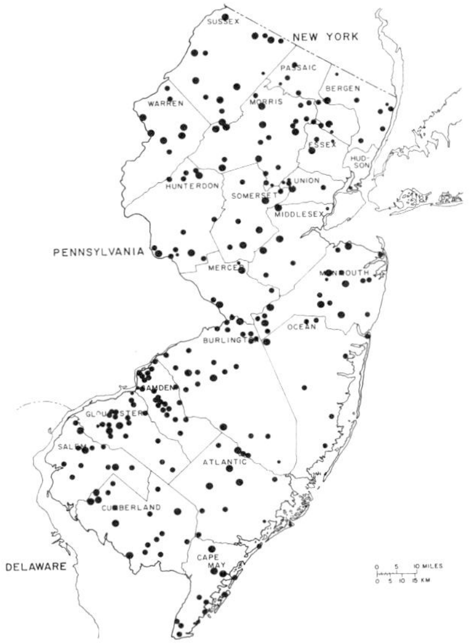

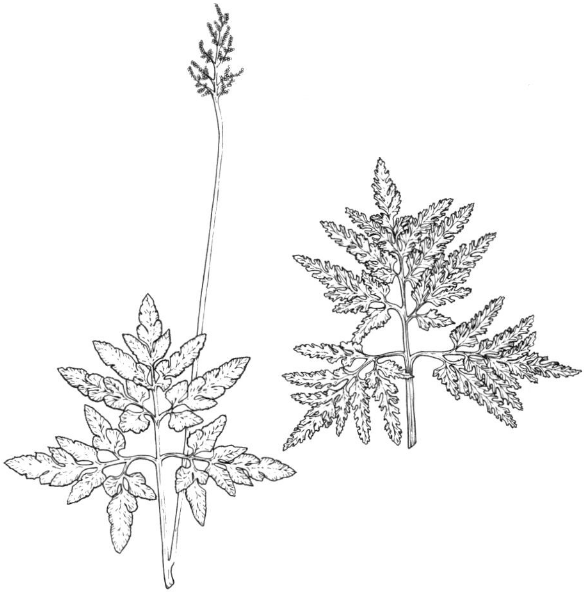

Botrychium multifidum (Gmel.) Rupr.

Leathery Grape Fern

Description: Fronds produced in midsummer, remaining green or turning bronze in winter, and lasting to the following summer; spores shed in late summer; vegetative blades at least tripinnate, triangular; segments rounded-serrate to entire, with obtuse apices; terminal segments nearly as broad as long and like the lateral segments in shape.

Habitat: Abandoned fields and pastures to open woods.

Range: Subarctic Canada, south to n NJ, w VA, IN, IL, and IA; various related taxa in w North America and Eurasia.

Distribution in New Jersey: Historical records in Sussex, Morris, Bergen, and Monmouth counties; recent records only in Warren County.

Comments: *Botrychium multifidum* was considered rare in New Jersey by Fairbrothers and Hough (1973), Snyder and Vivian (1981), and Montgomery (1989); and it was listed as critically imperiled in the state, although globally secure, by the New Jersey Natural Heritage Program (Snyder 1988). New Jersey is near the southern geographic limit, the range extending as far south as Shenandoah National Park, VA. Leathery grape fern is another successional species of abandoned fields, persisting into the shrub and young woodland stages. It does not usually occur in mature forests.

B. multifidum is diploid, with $2n = 90$.

Botrychium multifidum (Gmel.) Rupr.

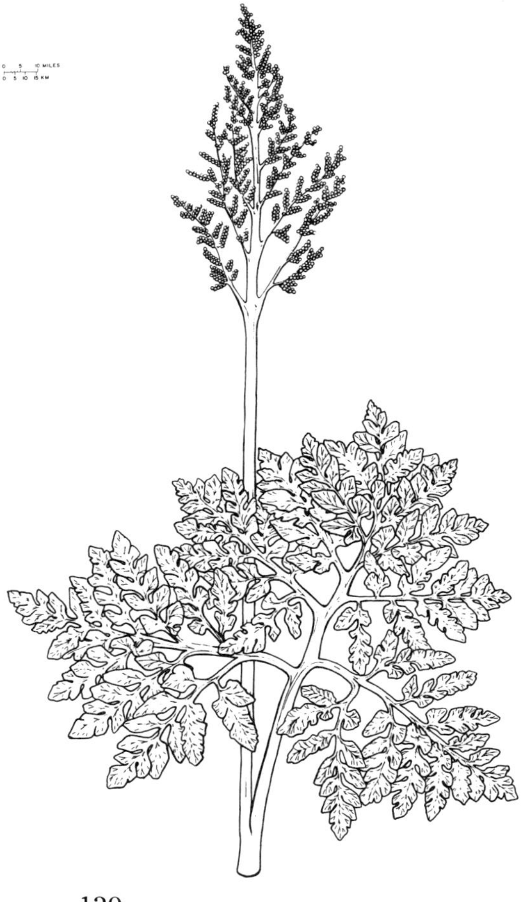

Botrychium oneidense (Gilbert) House

[*B. dissectum* var. *oneidense* (Gilbert) Waters]
[*B. multifidum* var. *oneidense* (Gilbert) Farw.]
Blunt-lobed Grape Fern

Description: Fronds produced in late summer, remaining green (not turning bronze) in winter, and lasting into the following summer; spores shed in late summer; vegetative blades bi- to tripinnate, triangular; segments rounded-serrate, with obtuse apices; terminal segments longer than broad, and more elongated than lateral segments.

Habitat: On alluvial soil in floodplain hardwood forests.

Range: PQ to n WI, south to n NJ, and mountains of VA, KY, and NC.

Distribution in New Jersey: Rare and widely scattered, from seven counties; recent records only in northern and central New Jersey.

Comments: *B. oneidense* was considered rare by Snyder and Vivian (1981) and Montgomery (1989), and listed as critically imperiled in the state as well as rare and local throughout its range by the New Jersey Natural Heritage Program (Snyder 1988). Confusion with *B. dissectum* has been part of the reason for its uncertain status in the state. The difference in habitat between blunt-lobed grape fern and other grape ferns is pronounced; there are no records of this species from abandoned fields, where other grape ferns are often found. *B. dissectum* occasionally grows in damp, second growth woods, especially in southern New Jersey, where *B. oneidense* does not occur, but rarely in mature floodplain forests.

B. oneidense is diploid, with $2n = 90$.

Botrychium oneidense (Gilbert) House

Ophioglossum

Small ferns of open habitats (epiphytic in the tropics). Rhizomes erect; plants producing a single fleshy leaf per season, divided into a simple lateral vegetative portion and an erect fertile portion with two rows of sporangia embedded in a fleshy axis. Two species occur in New Jersey.

Key to Species

1. Vegetative portion of the leaf broadest near the middle (elliptic), with no persistent basal sheath .... *O. pusillum*
1. Vegetative portion of the leaf broadest near the base (ovate), with a persistent basal sheath .... *O. vulgatum*

Ophioglossum pusillum Raf.

[*O. vulgatum* L. var. *pseudopodum* (Blake) Farw.]
Northern Adder's-tongue

Description: Vegetative blades elliptic, broadest near the middle and narrowing to the base, 2–10 cm long and 1–5 cm wide, dull green; sheath absent at base of stalk.

Habitat: Wet meadows, edges of swamps, and moist open woods.

Range: PQ to WA and AK, south to FL and uplands of VA, OH, IL, NE, and AZ.

Distribution in New Jersey: Scattered over the northern and southwestern counties, also in the Pinelands, but absent in the central counties.

Comments: This fern is more often overlooked than actually rare; it was considered rare in New Jersey by Fairbrothers and Hough (1972) and the New Jersey Natural Heritage Program (Snyder 1988).

In 1897, on a joint field excursion of the Torrey and Philadelphia Botanical Clubs, J. D. Crawford and C. L. Pollard found a colony of *Ophioglossum* from dry sandy soil in open woods between Holly Beach and Wildwood, Cape May County. This was described by E. G. Britton (1897) as *Ophioglossum arenarium*. The specimens have very

Ophioglossum pusillum Raf.

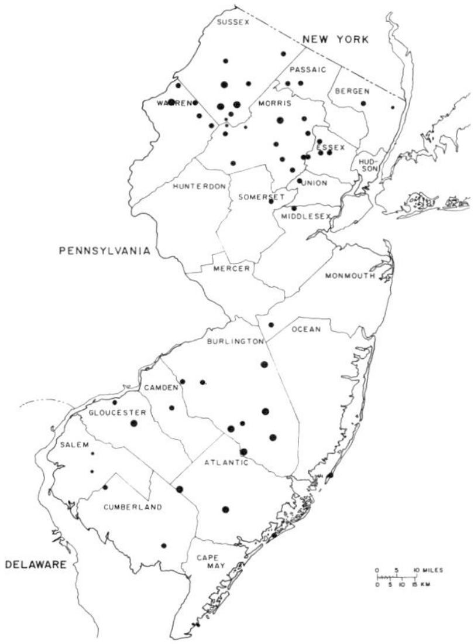

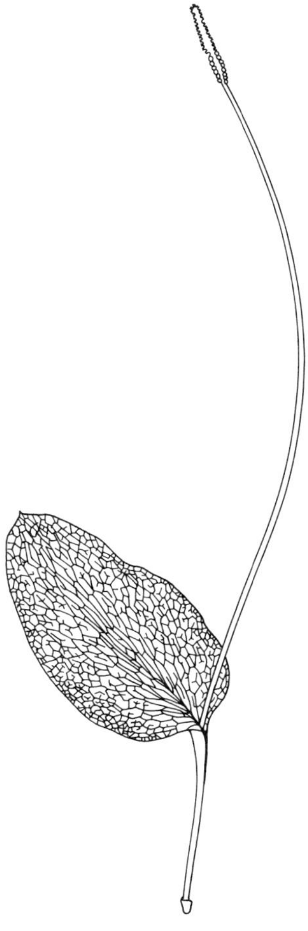

narrow, elongated vegetative blades. The status of this taxon remains unresolved, although it is usually considered a form or synonym of *O. pusillum*. It has not been collected since 1898, and Stone (1911) reported that the original colony had been destroyed. This taxon would be worth searching for in open woodlands near the coast in southern New Jersey.

Ophioglossum vulgatum L.

[*O. vulgatum* L. var. *pycnostichum* Fern.]
Southern Adder's-tongue

Description: Vegetative blade ovate, broadest near the base, 2–10 cm long, 1–5 cm wide, bright green; sheath persistent at the base of stipe.

Habitat: Moist loamy wooded slopes and creek banks.

Range: NJ, WV, and IN, south to FL and AL.

Distribution in New Jersey: Five records from Salem and Cumberland counties; last collected in 1958.

Comments: Since this taxon has been collected only once since 1950, it should be considered endangered, possibly extirpated, in New Jersey (Montgomery 1989). It was not listed by Snyder and Vivian (1981), but the New Jersey Natural Heritage Program listed it as having no known occurrence in the state (Snyder 1988).

The chromosome number of *O. vulgatum* is among the highest known for all organisms, with $2n = 1260$.

The two taxa of *Ophioglossum* in New Jersey are sufficiently different so that W. H. Wagner, presently revising this group, considers them distinct species. Many authors of recent fern books consider these as geographic varieties (Mickel 1979; Taylor 1984). The difference in color is usually evident in the field, but leaf shape is the best indicator for herbarium material, unless the base of the petiole has been collected to show the presence (*O. vulgatum*) or absence (*O. pusillum*) of the basal sheath.

Ophioglossum vulgatum L.

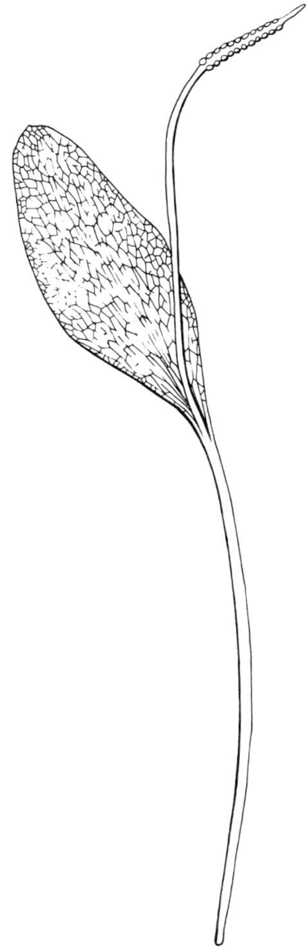

Osmunda

Large ferns of damp to wet habitats. Rhizomes erect, bearing a dense mass of fibrous roots; fronds deciduous, pinnate-pinnatifid to bipinnate; sporangia borne on fronds or pinnae without green tissue, separate from the vegetative fronds or pinnae; spores green.

Key to Species

1. Vegetative blades fully bipinnate; the fertile pinnae terminal on the blade .... *O. regalis*

1. Vegetative blades pinnate-pinnatifid; the fertile pinnae either on separate fronds or in the middle of the vegetative blades....................... **2**

  2. Tuft of hairs present in axils of vegetative pinnae; pinnae subacute at tip; fertile fronds distinct from vegetative fronds.... *O. cinnamomea*

  2. Tuft of hairs absent in axils of vegetative pinnae; pinnae rounded at the tip; fertile pinnae in middle of blade, among the vegetative pinnae .... *O. claytoniana*

Osmunda cinnamomea L.

Cinnamon Fern

Description: Fronds in a crown from a massive, upright rhizome; fertile and vegetative fronds distinct, the fertile appearing in the spring and withering soon after shedding spores; the vegetative 75–150 cm tall, with persistently wooly stipes; blades pinnate-pinnatifid, with subacute segments, a tuft of cinnamon-colored hair at the junction of the pinnae and rachis.

Habitat: Wet woods, swamps, open marshes, bogs, and roadside ditches.

Range: NF to MN, south to FL and TX; varieties in tropical America and Asia.

Distribution in New Jersey: Common throughout the state, with recent records from all 21 counties.

Comments: One of New Jersey's most common ferns, cinnamon fern covers large areas in marshes and wooded swamps in most parts of the state. In the Pinelands, where they grow together, *O. cinnamomea* and *Woodwardia virginica* are sometimes confused. The green

Osmunda cinnamomea L.

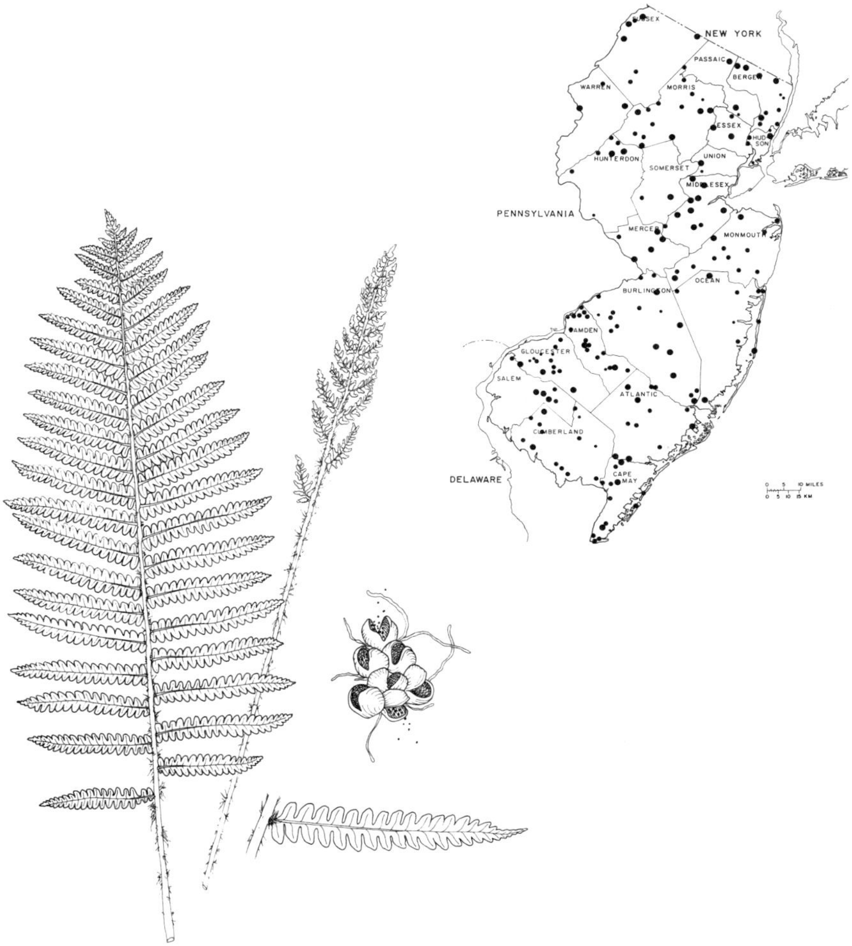

stipe, free veins, and axillary tuft of hair contrast with the black stipe, areolate veins, and lack of hair in *W. virginica*.

O. cinnamomea var. *glandulosa* was discovered by W. A. Poyser at Clementon, Camden County, in 1908. The leaf tissue appears to be darker and, when alive, has a brownish appearance owing to the presence of numerous glandular hairs on the underside of the lamina (W. H. Wagner, Jr., personal communication).

The massive roots have long been used for orchid culture, and the fiddleheads have apparently at times been used as food but are very bitter, hairy, and—in our opinion—unpleasant to eat.

The chromosome number of *O. cinnamomea* is $2n = 44$.

Osmunda claytoniana L.

Interrupted Fern

Description: Fronds in a crown from a massive rhizome; fertile pinnae 2–4 pairs in the middle of the blade, with vegetative pinnae above and below, the fertile pinnae withering soon after shedding spores in the spring, producing a gap in the blade; fronds 75–150 cm tall; stipes wooly when young, becoming glabrous; blades pinnate-pinnatifid with rounded segments, and lacking a tuft of hairs at the junction of the pinnae and rachis.

Habitat: Moist woods and edges of marshes and swamps; usually in areas that are not as wet as those of *O. cinnamomea*.

Range: NF to MN, south to GA, KY, and AR; a variety occurs in Asia.

Distribution in New Jersey: Common in the northern counties, becoming uncommon on the Inner Coastal Plain, and absent from the Outer Coastal Plain.

Comments: Of the three species of *Osmunda* in eastern North America, this is the best as a garden plant. Digging a mature plant, with its massive rhizome, is not advised; start with a small one! Although the cinnamon and interrupted ferns are often confused in the field (if no fertile pinnae are present), the lack of woolly tufts at the bases of the pinnae and the rounded shape of the segments of the interrupted fern are reliable characteristics. The fronds of the interrupted fern also tend to be more pea green in color, as opposed to yellow-green in the cinnamon fern. In the border of a swamp all three *Osmunda* species can be found, but the royal fern grows in the wettest areas, the cinnamon fern in those less wet, and the interrupted fern in habitats that are only damp.

The chromosome number of *O. claytoniana* is $2n = 44$.

Osmunda claytoniana L.

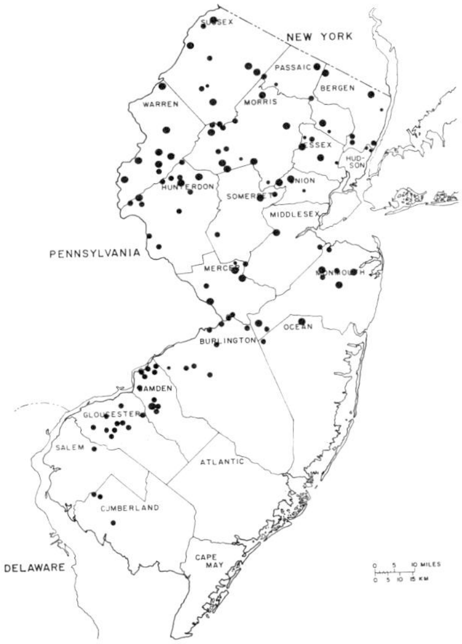

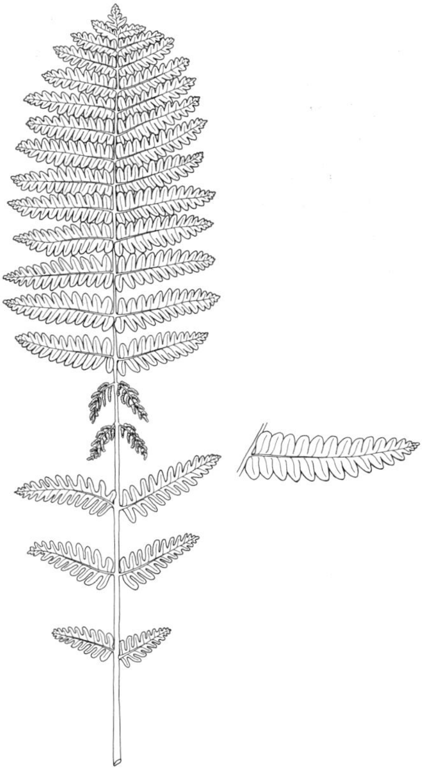

Osmunda regalis L. var. *spectabilis* (Willd.) Gray

Royal Fern

Description: Fronds in a crown from a massive rhizome; fronds up to 150 cm tall, stipes glabrous; fertile pinnae terminal, shedding spores in late spring and withering; vegetative blade bipinnate, with pinnules well separated, lance-eplliptic.

Habitat: Swampy woods, marshes, including borders of salt marshes, and bogs; cedar swamps in the Pinelands.

Range: NF to SK, south to FL, TX, and Mexico; also the West Indies and South America; var. *regalis* occurs in Europe, Asia, and Africa.

Distribution in New Jersey: Common in wet areas throughout the state, with recent records from all 21 counties.

Comments: Royal fern is one of our most distinctive species, with its bipinnate blades and swamp habitat. Although the three species of *Osmunda* often grow in the same area, a single hybrid has been detected, and found only twice (in CT and VA). This is *Osmunda* × *ruggii* Tryon, the hybrid between *O. regalis* and *O. claytoniana* (Wagner et al. 1978). The pinnae are pinnate but more ovate than in *O. regalis,* and the fertile pinnae may be apical or medial on the blade. Spores, as expected, are aborted. This hybrid could conceivably be found in New Jersey.

The chromosome number of *O. regalis* is $2n = 44$.

Osmunda regalis L.
var. *spectabilis* (Willd.) Gray

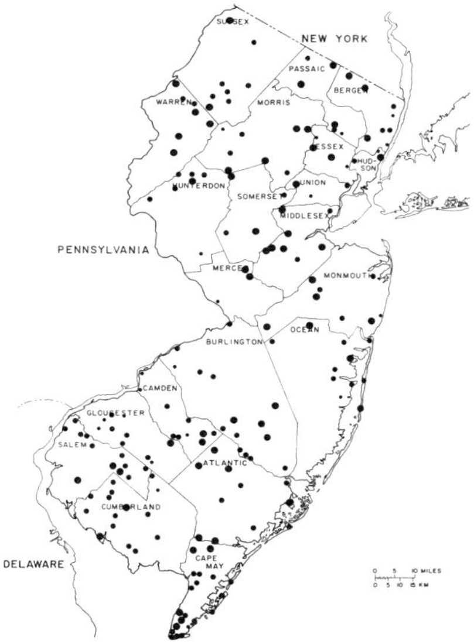

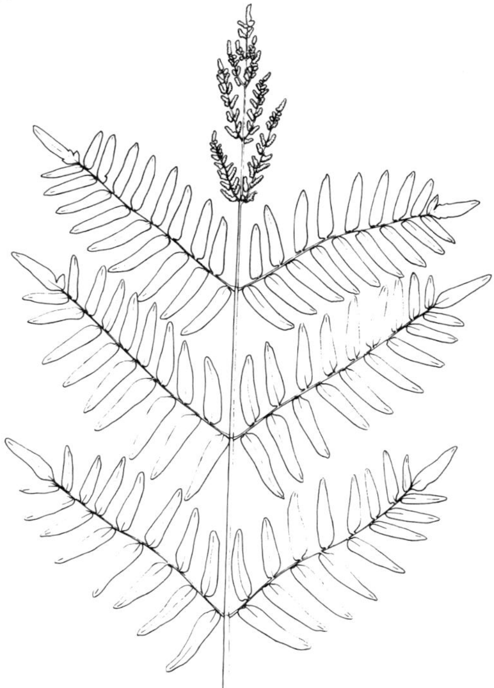

Schizaea

Small ferns of various terrestrial habitats, usually in sterile soil. Fertile and sterile segments on separate fronds (dimorphic) or terminal on the fronds; fronds linear to fan-shaped, simple to palmately lobed. Most of the species of *Schizaea* occur in the tropics or the Southern Hemisphere; the single species in New Jersey is isolated well north of other members of the genus.

Schizaea pusilla **Pursh**

Curly-grass Fern

Description: Fronds dimorphic, the vegetative fronds resembling small curled blades of grass, linear, up to 5 cm long; the fertile fronds erect, up to 13 cm long, with long stipe and pinnatifid blade with four to ten pairs of lobes, bearing sporangia on the lower surface.

Habitat: Wet acid bogs in the Pinelands. *Schizaea* occurs in the open bogs, not within dense white cedar forests. Plants are found at bases of young or isolated cedar trees, on stumps or logs, or on edges of peat hummocks including edges of old sand roads.

Range: Restricted to the Pinelands of NJ; e Long Island, NY; Sussex County, DE; and bogs in NS and NF.

Distribution in New Jersey: Pinelands of Monmouth, Ocean, Burlington, and Atlantic counties. with one or two stations in Camden, Gloucester, and Cumberland counties.

Comments: Curly-grass fern is unquestionably New Jersey's most famous plant. It was discovered in 1805 at Quaker Bridge, Burlington County, a stagecoach stop between Philadelphia and the Jersey coast, by C. W. Eddy, J. LeConte, F. T. Pursh, and C. Whitlow. Frederick Pursh described this species in his *Flora Americae Septentrionalis* of 1814. In this flora he cited collectors' names, specimens observed, or reference literature for included species described or cited by other botanists. These references were not included in the description statements if Pursh himself was the discoverer of the species; also, he rarely cited a herbarium specimen he had collected for the new species he described. Thus some controversy has arisen and Pursh has been justifiably criticized for creating the impression that he actually discovered *S. pusilla* at Quaker Bridge. Stone (1911) indicated that Dr. Eddy was the discoverer. This statement is based on a note published by J. H. Redfield (1869), who called attention to an annotation on specimens of *Schizaea* in the handwriting of Dr. John Torrey to the effect that the first specimens were discovered by Dr. Eddy, and "though [Dr. Eddy] and Mr. LeConte found all the specimens, Pursh has claimed the honor of the discovery himself." A note by W. Cooper indicated that the "first specimens were found in 1805, and not found again till detected by me in company with Dr. Torrey, in June, 1818." The 1818 excursion was John Torrey's first to the Pinelands, recounted in his letter to Zaccheus Collins in July 1818 (Redfield 1876).

The original herbarium specimens apparently have been lost. The earliest extant collection may be a herbarium sheet in the Schweinitz Herbarium at the Academy of Natural Sciences of Philadelphia. It includes individual plants collected by one of the Steinhauer brothers, who came to this country in 1815; and collections by Frederick Kampman, a naturalist from the Moravian settlement at Hope, NJ. There are specimens collected at Quaker Bridge in June 1818 by Cooper, Green, and Boott in both the Link Herbarium at the Berlin–Dahlem Museum and at the British Museum—Natural History in London. Additional later collections were made from Quaker Bridge, and this remained the only locality known until it was collected in 1834 by John Torrey at Toms River, NJ. Localities then increase steadily: Batsto, 1861; Pleasant Mills, 1861; Little Egg Harbor, 1865; etc. Stone (1911) reported 30 stations. Counting sites becomes more or less meaningless, since *Schizaea* occurs in scattered populations in the open bogs that border streams, such as the Batsto and Oswego, for several miles through the Pinelands.

Curly-grass fern is notoriously difficult to find, and everyone has a favorite story about it being discovered while eating lunch, collecting other plants, and the like. Knowing where to search is important. This fern grows in the open bogs and the edges of hummocks. It does not grow in the dense shade of white cedar forests, or in dense mats of *Sphagnum*. The best indicator associates are thread-leaf sundew (*Drosera filiformis*), and Carolina clubmoss (*Lycopodium carolinianum*). Knowing when to look is also helpful. The fern is most easily seen in late fall and early winter, even when there is a light cover of snow. At this time the small, bright green vegetative fronds and the cinnamon-brown fertile fronds contrast with dead grasses and sedges in the bogs.

Dr. G. H. E. Muhlenberg listed the name *Schizaea tortuosa* for this species in his catalog published in 1813 (and in the second edition of 1818). This is an invalid name because there is no description (nomen nudum). Bachelot de la Pylaie, a French botanist, described

Schizaea filifolia from Newfoundland in 1826 (p. 424), not realizing that it was a later name for the same species as the recently described *S. pusilla* from New Jersey. There are several locations in Newfoundland and Nova Scotia, Canada. A finding of curly-grass fern at a locality near Montauk, Long Island, was the first published report from New York State (Moldenke 1960); however, there are herbarium specimens at the New York State Museum in Albany collected by R. Latham in 1927–1929 that represent either the same or a nearby population. Two populations are presently known on Long Island. *S. pusilla* has recently been reported from Sussex County, Delaware (Hirst 1990); we have not seen specimens to verify this report. A report of *S. pusilla* from Ontario, Canada (H. H. Brown 1945), has been discounted by most recent authors (Wherry 1961; Stolze 1987), since this is a limestone area; also there may have been confusion about the location, since the collections were stored for a long time. On the other hand, *S. pusilla* has recently been reported from Peru (Stolze 1987), an enormous geographic disjunction. Even this was discovered in the typical way, when the collector slipped while crossing a landslide!

S. pusilla was listed as local in New Jersey by Snyder and Vivian (1981) and Snyder (1984), and was considered rare but not presently imperiled in the state by the New Jersey Natural Heritage Program (Snyder 1988). The species is listed as a protected plant by the Pinelands Commission. It has been proposed for listing by the U.S. Fish and Wildlife Service but is not presently under federal review as endangered or threatened. While the species is not rare in New Jersey, it is not found elsewhere in the United States except for the populations on Long Island and Delaware. Thus protection of *S. pusilla* remains of national, rather than strictly local, interest (Montgomery 1989).

The life history and anatomy of *S. pusilla* were first described in detail by E. G. Britton and A. Taylor (1901). The gametophyte is filamentous and green, bearing special spherical branches that contain an endophytic (symbiotic) fungus in contact with the substrate. Gametophytes may be found on damp sand, decayed *Sphagnum,* or old logs. Atkinson (1965) reported that the gametophytes can survive cold temperatures and live for several years if fertilization does not occur. Spores mature in autumn and are shed by early winter. They apparently remain dormant until spring; Guiragossian and Koning (1986) have shown that spore germination requires light and is increased by a low-temperature pretreatment in the dark.

The chromosome number of *S. pusilla,* reported by Wagner (1963), is $2n = 206$.

Schizaea pusilla Pursh

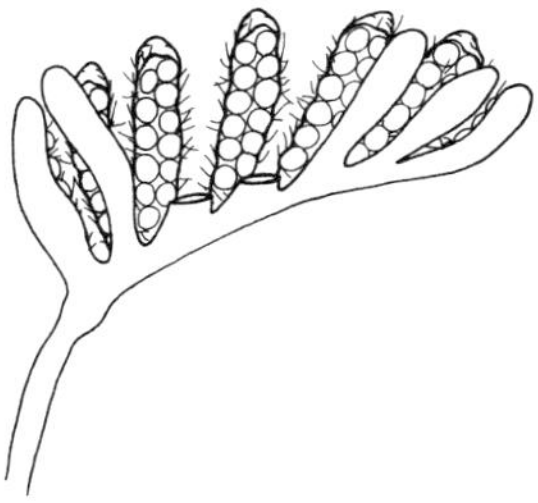

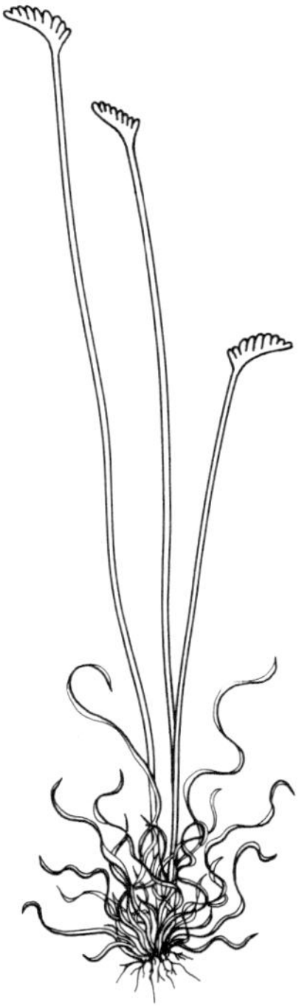

Lygodium

Ferns with climbing fronds. Blades linear, with palmately or pinnately branched pinnae; fertile areas or segments terminal, sori pocket-like structures with few sporangia. The genus is widely geographically distributed; a single species occurs in New Jersey.

Lygodium palmatum (Bernh.) Sw.

Climbing Fern, Hartford Fern

Description: Rhizome long-creeping; fronds linear, climbing by twining rachis on other vegetation, up to 2 m long; vegetative portion below, twice pinnate, with pinnae divided into two palmately lobed pinnules; fertile portion above, dichotomously forked into palmately lobed segments.

Habitat; Usually near streams in thickets, edges of bogs, and swampy open woods, with the soil intensely acid.

Range: s NH to NY and OH, south to WV, KY, TN, and GA; rare and scattered throughout most of the range except KY and TN.

Distribution in New Jersey: A few stations on the western edge of the Pinelands; earlier records from the central and southwestern part of the state.

Comments: The earliest record with specific locality for the state is from New Lisbon, Burlington County, in 1842 (an 1840 collection by Asa Gray gives no specific locality within the state). There are several other records from the edge of the Pinelands from 1845–1870 (Shark River, Mount Holly, Browns Mills). Some of these stations are still extant. There is also a series of records from Middlesex and Mercer counties from the mid-nineteenth century up to 1938. None of these stations are extant, although climbing fern was reported as abundant in the early twentieth century (House 1904).

L. palmatum was listed as endangered in New Jersey by Fairbrothers and Hough (1973), Snyder and Vivian (1981), Snyder (1984), and Montgomery (1989). It was listed as imperiled in the state and apparently globally secure although rare in parts of its range by the New Jersey Natural Heritage Program (Snyder 1988). In addition this species was reported as rare or endangered in at least 12 state lists in the northeastern United States (Montgomery 1989). *L. palmatum*

Lygodium palmatum (Bernh.) Sw.

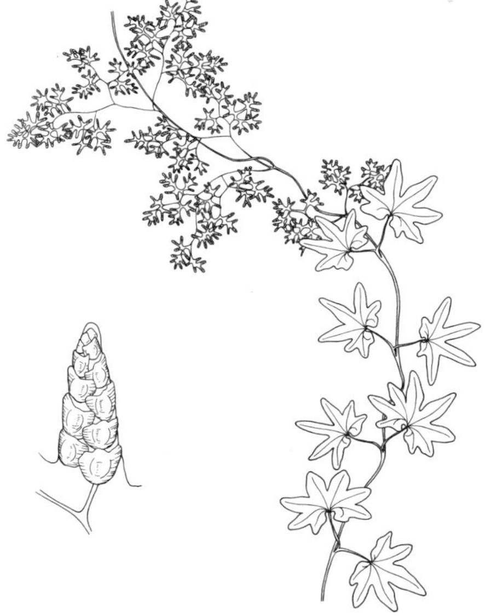

seems always to have been rare, but it represents a fairly typical case of a plant from a restricted habitat (acid wetlands) that has been subject to much destruction since World War II. In a rare case, possibly unique for a fern, a highway in Burlington County was recently rerouted to avoid destruction of one of the few remaining populations of this fern in the state (Montgomery 1989). The legislature of Connecticut in 1869 passed a special law for the protection of *L. palmatum*. In 1875 the law was codified, stating that the offense of picking this fern species was punishable by a fine not to exceed $100.00 or imprisonment not more than 12 months, or both. Prof. D. C. Eaton (1879) indicated that it was the first time a statute in the United States designated a plant to receive special legal protection solely on account of its beauty.

The chromosome number of *L. palmatum* is $2n = 60$.

Adiantum

Medium-to-large ferns, usually of forest habitats or on rocks. Rhizome creeping; fronds very diverse in form, from simple to quadripinnate; stipes often shiny purple or black; sori marginal, elongate, borne on the under side of the false indusium.

Adiantum is a large genus, primarily in the tropics, but widely distributed. Many species are cultivated for their showy foliage. A single species occurs in Jersey.

Adiantum pedatum L.

Maidenhair Fern

Description: Fronds all alike, deciduous, up to 50 cm tall; with distinctive architecture; stipes shiny black, divided into two branches, from which arise secondary branches, bearing the oblong ultimate segments; sori oblong on the margins of the segments.

Habitat: Mesic woods, usually in rich soil.

Range: NS to MN and SD, south to GA, LA, and OK.

Distribution in New Jersey: Common, especially in the limestone area in northern New Jersey; less common southward, and only one recent record in southern New Jersey.

Comments: As noted by Chrysler and Edwards (1947), this is the one New Jersey fern that everyone recognizes. It is readily cultivated in a moist woodland garden, spreading fairly rapidly. The name maidenhair probably refers to the drooping pinnae.

Recent research indicates that there are three taxa of maidenhair ferns in the northeastern United States: a widespread diploid ($2n = 58$) growing on various soil types; a rare diploid on serpentine soils; and a very rare tetraploid on serpentine soils (Paris and Windham 1988). Only the widespread diploid has been collected in New Jersey.

Adiantum pedatum L.

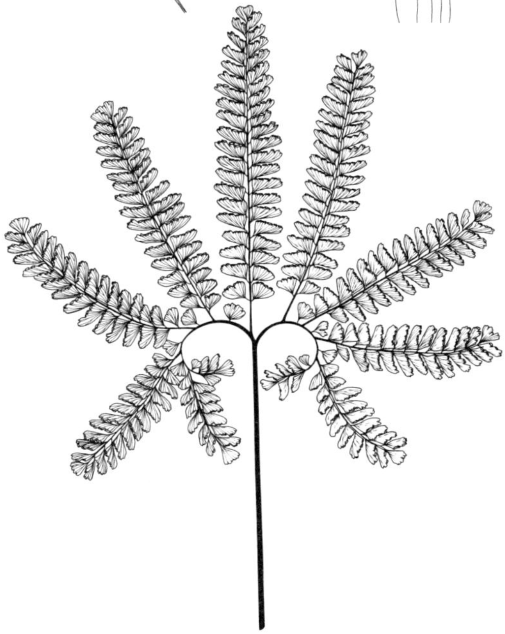

Cryptogramma

Small rock ferns. Rhizome creeping; fronds dimorphic, the fertile narrower and longer than the vegetative; sorus continuous along the margin of the fertile segments, with false indusium. A small genus, with a single species in New Jersey.

Cryptogramma stelleri (Gmel.) Prantl

Slender Rock-brake

Description: Fronds bipinnate, deciduous, the vegetative lanceolate-ovate, up to 10 cm long, with rounded pinnae; the fertile lanceolate, up to 25 cm long, with narrow, subacute pinnae; sori continuous, with the margin folded over as a false indusium.

Habitat: Cliffs and ledges, mostly on calcareous rocks.

Range: Labrador to AK, south to n NJ, n PA, WV, MI, CO, and WA; also in Asia.

Distribution in New Jersey: Extirpated; the single station was in Bergen County.

Comments: *C. stelleri* was first collected by C. F. Austin in 1869, from "Godwinsville," usually called Wortendyke, now part of Midland Park, in Bergen County. It was not collected again until 1917, when the station was rediscovered. There are collections until about 1950. A search by Montgomery in 1962 and by others subsequently have found no plants at the site. The severe drought of the early 1960s may have eliminated the plants (Montgomery 1989). New Jersey was at the southern geographic limit of this boreal species. The habitat was somewhat unusual in being conglomerate, but some lime may be present in the matrix (Chrysler and Edwards 1947).

The chromosome number of *C. stelleri* is $2n = 60$.

Cryptogramma stelleri (Gmel.) Prantl

Cheilanthes

Small to medium-sized ferns of dry, rocky habitats; fronds all alike, usually hairy or scaly on the blade; sori nearly round, marginal, covered or partly so by false indusia. A fairly large genus, especially of the southwestern United States and Mexico; a single species occurs in New Jersey.

Chielanthes lanosa (Michx.) D.C. Eaton

Hairy Lip Fern

Description: Rhizomes short-creeping; fronds lance-oblong, evergreen, bipinnate-pinnatifid, 15–30 cm tall, blades with jointed hairs below; rachis chestnut brown; ultimate segments oblong, lobed, with margins rolled under to form false indusia over the sori.

Habitat: On rock ledges of sandstone and traprock.

Range: CT, n NJ, MI, and MN, south to MO, KS, TX, and VA.

Distribution in New Jersey: Presently known from the Delaware River cliffs in Hunterdon and Sussex counties; older records from northeastern New Jersey.

Comments: *C. lanosa* was listed as endangered by Fairbrothers and Hough (1973), rare by Snyder and Vivian (1981) and Montgomery (1989), and imperiled in the state by the New Jersey Natural Heritage Program (Snyder 1988). There were several stations in the Palisades and Watchungs of eastern New Jersey, including Snake Hill, in Hudson County, where it was collected by F. J. Bumstead in 1865 (Eaton 1879); it is now restricted to a few rock outcrops along the Delaware River. Montgomery (1989) speculated that not only habitat destruction but also air pollution may have eliminated rock ferns from the northeastern counties.

The chromosome number of *C. lanosa* is $2n = 60$.

Cheilanthes lanosa (Michx.) D. C. Eaton

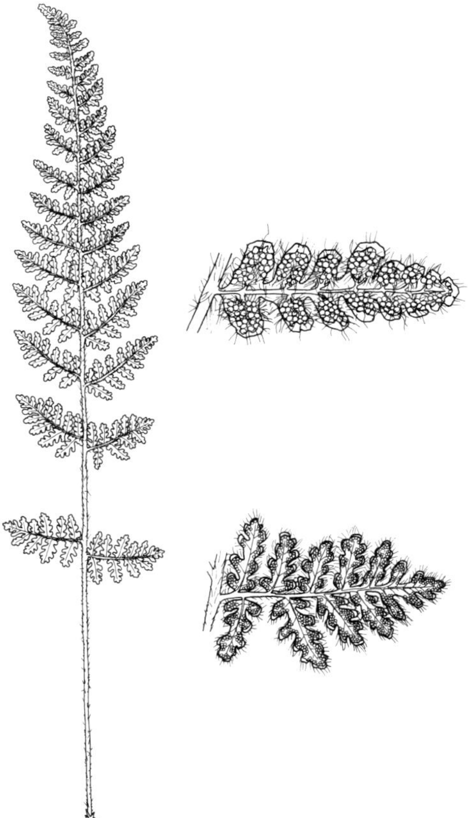

Pellaea

Small to medium-sized ferns, mostly of rocky habitats. Fronds pinnate to bipinnate, with glabrous ultimate segments; sori marginal in a continuous line, with slight or no false indusia, but the segments often rolling lengthwise, which reduces dessication during dry periods. *Pellaea* is a worldwide genus.

Key to Species

1. Stipes shining brown, glabrous except for a few scales at the base; fertile and vegetative fronds alike .... *P. glabella*
1. Stipes dull brown or purple, rough hairy; fertile fronds larger than vegetative fronds ....*P. atropurpurea*

Pellaea glabella Mett. ex Kuhn

Smooth Cliff-brake

Description: Rhizome short-creeping; fronds all alike, evergreen, up to 20 cm long, pinnate to bipinnate at base; stipes shining brown or purple, glabrous; blade blue-green, segments oblong.

Habitat: Exposed cliffs and ledges of limestone; also on masonry of bridges.

Range: PQ to BC, south to VA, TN, AK, TX, and CO.

Distribution in New Jersey: Uncommon, mostly from calcareous rocks in Sussex and Warren counties, as well as a few stations on bridge abutments in Sussex and Hunterdon counties.

Comments: This species is fairly tolerant of exposed conditions, and several of the New Jersey stations are on bridges. In Frances T. Parsons's early fern book (1899), smooth cliff-brake was described as "one of the plants that rejoice in un-get-at-able and perilous situations."

Pellaea glabella is tetraploid and asexual, with $2n = 116$.

Pellaea glabella Mett. ex Kuhn

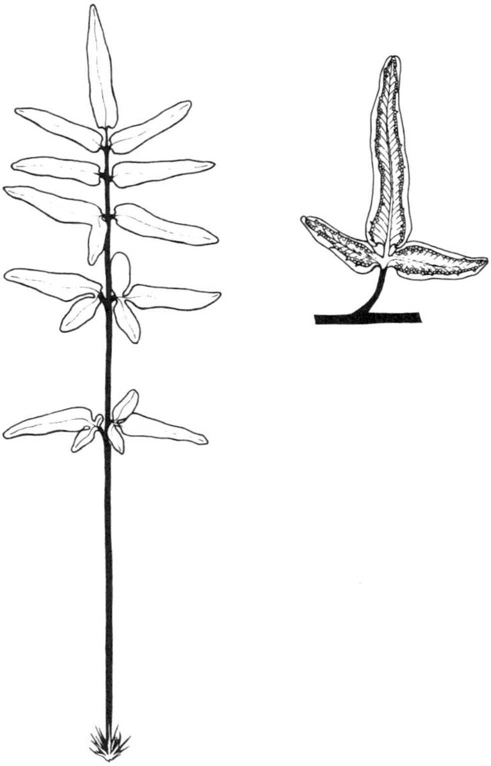

Pellaea atropurpurea (L.) Link

Purple Cliff-brake

Description: Rhizome short-creeping; fronds evergreen, blue-green in color, pinnate above to bipinnate at base; fronds somewhat dimorphic, the fertile up to 40 cm tall and taller than the vegetative; stipes dull brown or purple, with short hairs; ultimate segments oblong.

Habitat: Crevices in cliffs and ledges of limestone or occasionally traprock.

Range: VT and n NY to SD, south to AZ and FL; also in Mexico and Guatemala.

Distribution in New Jersey: Common in the limestone area of Sussex and Warren counties; a few other stations on traprock (Morris County).

Comments: This species, like the smooth cliff-brake, will grow on old masonry, but requires less exposed conditions. *P. atropurpurea* is triploid, with $3n = 87$. However, *P. atropurpurea* has an apogamous life cycle, that is, it produces sporophyte plants directly from the gametophyte without fertilization. Spores are produced that appear normal, although there are only 32 per sporangium. In this way a triploid, usually expected to be sterile, has become functionally fertile and acts as a distinct species.

Pellaea atropurpurea (L.) Link

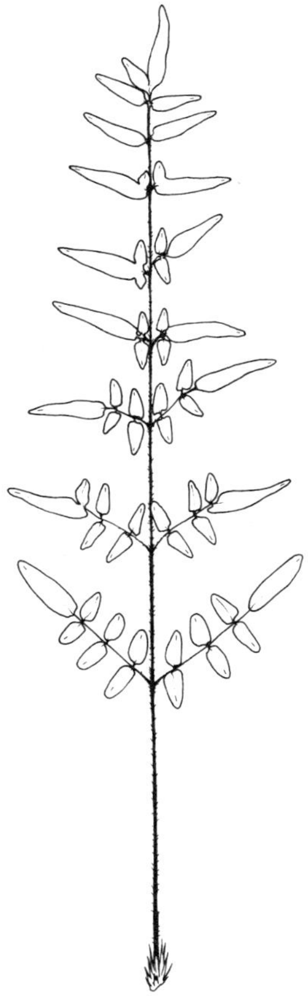

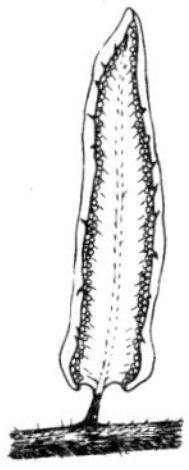

Trichomanes

Epiphytic ferns in the tropics. This genus is represented by two species in the southeastern United States. A single species occurs only as a gametophyte in New Jersey.

Trichomanes sp.

Appalachian Gametophyte

Description: Known only as a gametophyte. Appears as a dense green colony of filaments with small vegetative buds (gemmae).

Habitat: Deep under overhanging ledges and caves of noncalcareous rock.

Range: NH, VT, NY, PA, and OH, south to GA and AK.

Distribution in New Jersey: Presently known from a single station in Warren County.

Comments: This peculiar plant was first discovered in New Jersey in 1981 (Farrar et al. 1983). Colonies occur only in the darkest recesses of overhanging rocks in canyons and gorges where there is a constant moisture supply. The plants look much more like algae than ferns, but the small buds (gemmae) are distinctive.

The gametophytes can spread by these gemmae. Sex organs have not been found, and no sporophytes are formed. This makes identification using the usual morphological characteristics impossible. The plants may be a specialized form of one of the southeastern species or may be a relict species persisting only as a gametophyte. Farrar, who has studied these ferns extensively (see Farrar et al. 1983), has postulated that *Trichomanes* spread northward during warmer postglacial periods, possibly as sporophytes, but only the gametophytes can tolerate the present climate.

Polypodium

Small to medium-sized ferns, mostly tropical epiphytes, with a few temperate species. Rhizomes long-creeping; fronds pinnatifid to pinnate, rarely entire to bipinnatifid; sori round to oval, dorsal, with no indusia. *Polypodium* and several related genera constitute the most common epiphytic ferns of the American tropics. The temperate species are mostly small and terrestrial. A single species occurs in New Jersey.

Polypodium virginianum L.
Polypodium appalachianum Haufler × Windham

Common Polypody, Rock-cap

Description: Rhizomes long-creeping; fronds all alike, evergreen, 10–25 cm tall, lanceolate, deeply pinnatifid (not fully pinnate); segments usually round at the apex; sori in two rows on each segment, orange-brown.

Habitat: Boulders or ledges in moist to damp woods, less often on soil or tree bases.

Range: NF to ne BC, south to VA and uplands of GA, AL, TN, and AR.

Distribution in New Jersey: Common in northern New Jersey, more scattered on the Inner Coastal Plain, and a few stations in the Pinelands, recorded from all 21 counties, but few recent records from the central and southern part of the state.

Comments: The name rock-cap is certainly appropriate, especially in northern New Jersey, where this fern can be found covering large boulders in the woods (especially hemlock stands).

Several varieties and forms have been described but are presently not considered worthy of taxonomic recognition.

P. virginianum has been found to include diploid ($2n = 74$), triploid ($3n = 111$), and tetraploid ($2n = 148$) races (Kott and Britton 1982, Haufler and Windham 1991). The diploid, recently named *P. appalachianum*, has wider, more deltoid blades, with lobed tip and more acute segment apices. Tetraploids, *P. virginianum*, have narrower, ovate-lanceolate blades, with unlobed apex and more rounded segment apices. The triploid is more or less intermediate in these features but can be best distinguished by the mostly abortive spores. *P. appalachianum* and *P. virginianum* both occur in New Jersey; the triploid also probably occurs in the state. The publication of this information took place too recently for separation of species records in New Jersey; therefore, the map includes both taxa.

Polypodium virginianum L.

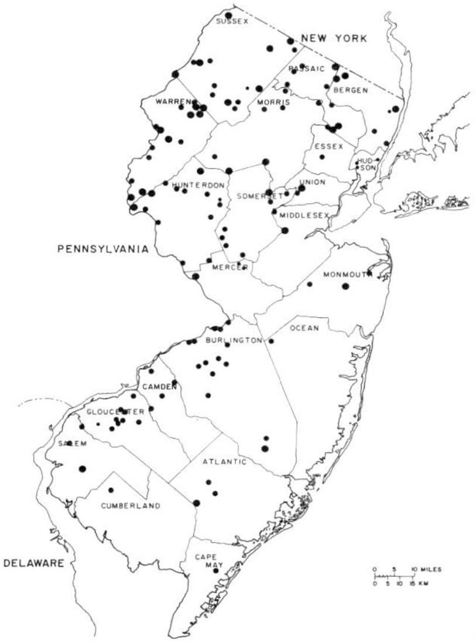

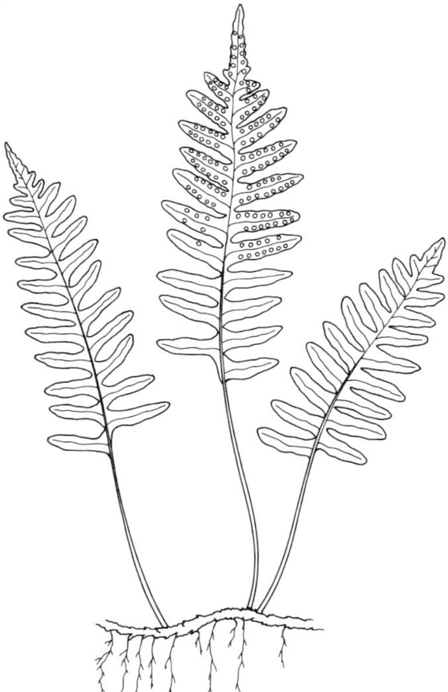

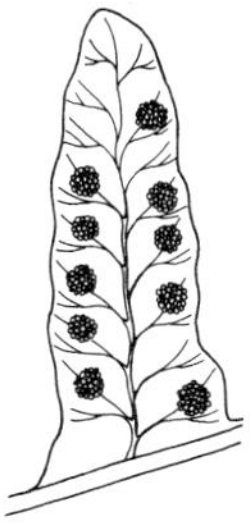

Dennstaedtia

Large terrestrial ferns of forests or forest borders. Rhizomes long-creeping; fronds with finely divided blades, hairy, but not scaly; sori round, marginal, borne in cup-like indusia. Most species are tropical, and the genus contains only one species in the temperate United States.

Dennstaedtia punctilobula (Michx.) Moore

Hay-scented Fern

Description: Rhizomes long-creeping, the plants forming extensive colonies; fronds up to 120 cm tall; stipes hairy; blades deciduous, bi- to tripinnate, with glandular hairs on the surface.

Habitat: Open banks and clearings in dry to mesic woods, also on roadside and railroad banks; often forming extensive colonies in partial or full sun.

Range: NF and sw PQ to MI and MO, south to GA and AR.

Distribution in New Jersey: Abundant in the northern counties, becoming less common southward, uncommon in the Pinelands; recent records for all counties except Cape May.

Comments: Hay-scented fern is a familiar plant of roadsides, especially along rocky walls in northern New Jersey. The common name refers to the odor of crushed fronds when fresh. Although this fern is attractive, it is too invasive for most garden uses, but it can be introduced onto roadside banks for erosion control.

The chromosome number of *D. punctilobula* is $2n = 68$.

Dennstaedtia punctilobula (Michx.) Moore

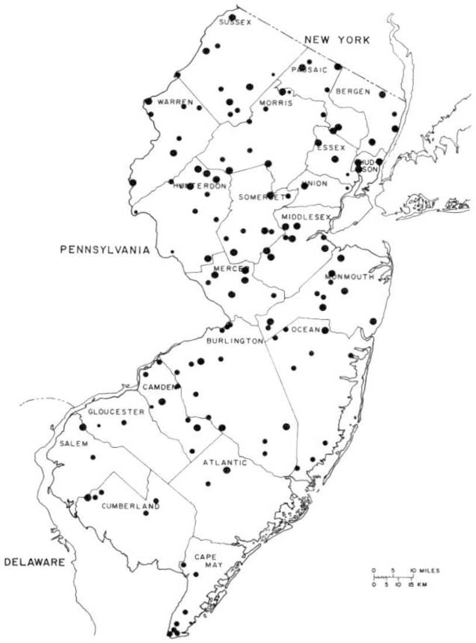

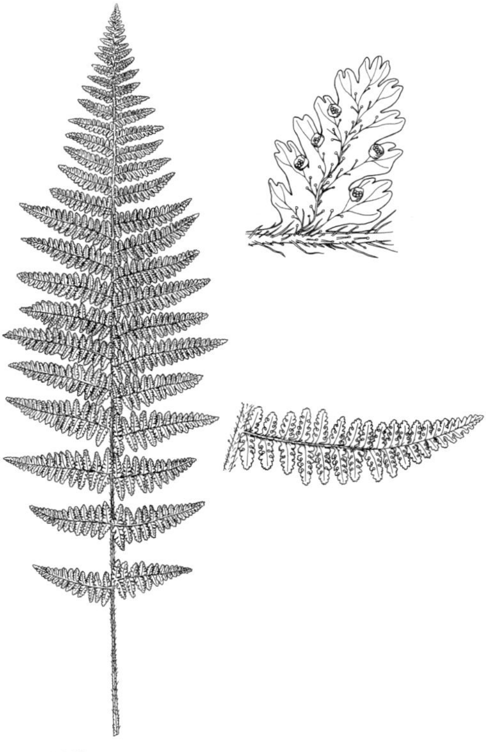

Pteridium

Large aggressive ferns of open habitats. Rhizome long-creeping, hairy, but not scaly; stipes stout; blades large and dissected; sori marginal, continuous, with inner true indusia and outer false indusia formed from the underroled margin of the segments.

Pteridium is a genus of worldwide distribution, composed of a single highly variable species or several closely related ones. Until a modern systematic treatment is prepared, the New Jersey taxa can be treated as a single species with two varieties.

Key to Varieties

1. Terminal segment 2–4 times longer than wide, 5–8 mm wide . . . . *P. aquilinum* var. *latiusculum*
1. Terminal segment 5–10 (or more) times longer than wide, 3–5 mm wide . . . . *P. aquilinum* var. *pseudocaudatum*

Pteridium aquilinum (L.) Kuhn var. *latiusculum* (Desv.) Underw.

Northeastern Bracken

Description: Rhizome long-creeping, forming extensive colonies; fronds up to 150 cm tall; blades triangular-ovate, tripinnate-pinnatifid, the terminal segments two to four times longer than wide, the lower surface short-hairy to glabrous.

Habitat: Dry open woods, open fields, and borders of wetlands, usually in partial to full sun; in the Pinelands in pine-oak woods.

Range: NF to MN, south to NJ, PA, uplands of NC, TN, MO, and OK; also in Mexico, Europe, and e Asia.

Distribution in New Jersey: Common to abundant throughout the state, with recent records for all counties except Hudson.

Comments: Bracken is known to be poisonous to cattle and carcinogenic to humans (Jarrett 1982), although it continues to be eaten in the Orient. Bracken readily invades pastures, especially when they have been overgrazed. For this reason, it is of considerable economic

Pteridium aquilinum (L.) Kuhn
var. *latiusculum* (Desv.) Underw.

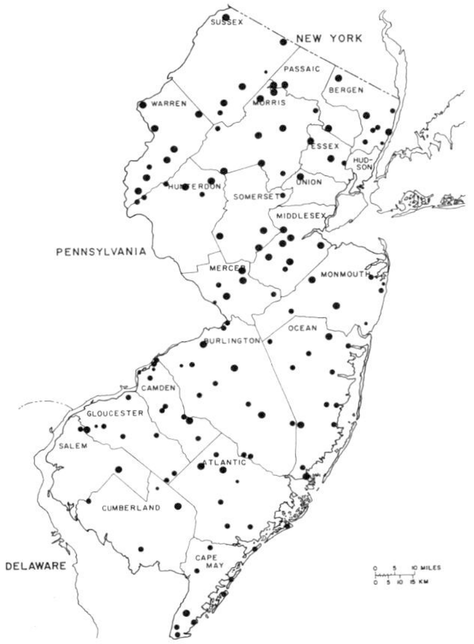

importance, especially in Europe, and has been the subject of much research (Kirkwood and Hinshalwood 1985). Bracken is frequently the major ground cover in the Pinelands following fires and may occur in extensive areas. It tends to decrease as the forest canopy increases, but the rhizomes persist with a few scattered fronds until the next fire opens the canopy again.

The chromosome number of *P. aquilinum* is $2n = 104$.

Pteridium aquilinum (L.) Kuhn var. *pseudocaudatum* (Clute) Heller

Southeastern Bracken

Description: Much like the northeastern bracken, but the ultimate segments much elongated, five to ten times longer than wide, and narrower, only 3–5 mm wide; blades glabrous below.

Habitat: Damp to dry open woods, either hardwood or mixed pine-hardwood.

Range: se MA and s OH to e OK, TX, and FL.

Distribution in New Jersey: Scattered in the Pinelands and in Cape May County.

Comments: The northeastern and southeastern bracken seem to intergrade in southern New Jersey. Since a single bracken plant can cover a large surface area, it is difficult to determine what constitutes a single clone. In the Pinelands individual fronds can be collected that are typical of one variety or the other, but some are more or less intermediate in morphology between the two. A study of the morphological variation, geographic ranges, and biosystematics of the two taxa is needed. Both varieties of bracken are often sterile in New Jersey, with indusia indicating position of sori, but no sporangia present. Thus it has been impossible to determine whether hybrids occur or are fertile.

Pteridium aquilinum (L.) Kuhn
var. *pseudocaudatum* (Clute) Heller

Thelypteris

Medium-sized woodland or wetland ferns. Rhizome creeping, plants therefore forming colonies. Fronds deciduous; stipes with two vascular bundles; stipes, rachises, and minor axes with distinctive, one-celled pointed hairs; sori round, with reniform indusia or indusia absent. *Thelypteris* resembles *Dryopteris* in gross form, sori, and in habitat requirements, but the two genera are placed in separate families in recent classifications. The hairs on the axes (*Dryopteris* has scales or glandular hairs), two vascular bundles in the stipes, and the round sori will distinguish *Thelypteris* from other woodland and swamp ferns.

Thelypteris, if recognized as a single genus, contains more than 1000 species, mostly in the tropics of both hemispheres. The temperate species of New Jersey are representatives of three taxonomic groups, as indicated by Lellinger (1985), or three separate genera in the classification of Holttum (1971). Chromosome numbers differ between these groups and are one of the lines of evidence supporting separation of genera (Holttum 1983).

Key to species

1. Blades triangular in outline, nearly as broad as long, pinnatifid, but not compound except for the basal pinna pair; indusia absent **2**

2. Basal pinna pair separate, not connected by a wing along the rachis; blades longer than wide .... *T. phegopteris*

2. All pinnae adnate by a wing along the rachis; blades nearly as wide or wider than long.................................. *T. hexagonoptera*

1. Blades elliptic to lanceolate in outline, 2–4 times longer than broad; blades pinnate-pinnatifid; indusia present **3**

3. Blades tapering to base, lowest pinnae reduced to small wings......... .... *T. noveboracensis*

3. Blades truncate at the base, lowest pinnae not much reduced........ **4**

4. Veins in nonfertile segments not forked .... *T. simulata*

4. Veins in nonfertile segments once forked .... *T. palustris*

Thelypteris hexagonoptera (Michx.) Weatherby

[*Phegopteris hexagonoptera* (Michx.) Fee]
Southern Beech Fern, Broad Beech Fern

Description: Fronds broadly triangular, up to 50 cm long, blades as wide or wider than long; pinnatifid, but not fully pinnate, and all segments joined by a wing along the rachis, with larger segments also lobed; sori lacking indusia.

Habitat: Mesic to damp, rich woods.

Range: PQ to MN, south to FL, TX, and KS.

Distribution in New Jersey: Common in northern New Jersey, and scattered in the counties bordering the Delaware River.

Comments: This species is much more common in New Jersey than the Northern Beech Fern. It is a good choice for a woodland garden because it is attractive, fairly easy to grow, and spreads slowly by creeping rhizomes.

There are three species of beech ferns—the two in this book and *T. decursive-pinnata* (van Hall) Ching, from Japan, which looks more like our New York fern. These three species are sometimes segregated as a genus, *Phegopteris*. Our treatment follows those of Morton (1963) and Smith (1971) in maintaining the inclusive genus *Thelypteris*.

T. hexagonoptera is diploid, with $2n = 60$.

Thelypteris hexagonoptera (Michx.) Weatherby

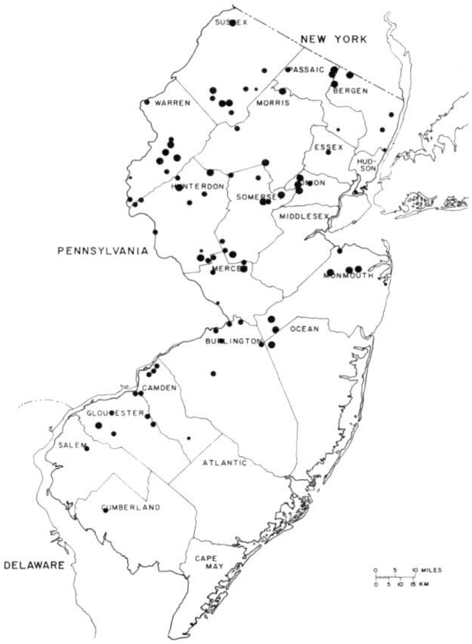

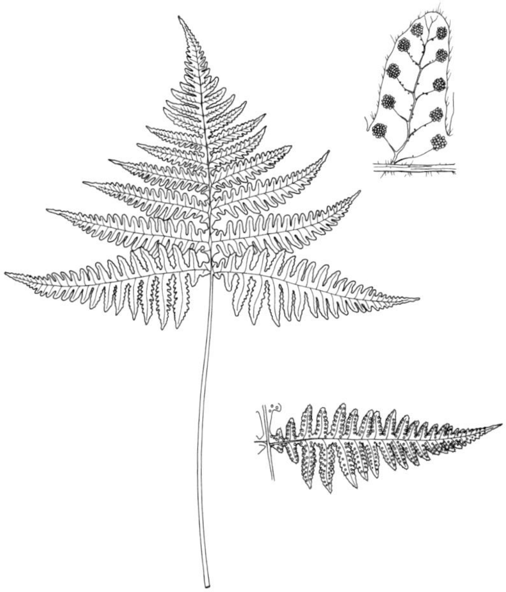

Thelypteris phegopteris (L.) Slosson

[*Phegopteris polypodioides* Fee]
[*P. connectilis* (Michx.) Watt.]
Northern Beech Fern, Long Beech Fern

Description: Fronds triangular, up to 40 cm long; blades longer than wide, basal pinna pair separate (pinnate), segments above the basal pair connected by wings along the rachis, larger ones lobed; sori lacking indusia.

Habitat: Cool moist ravines or edges of swamps.

Range: Greenland and Labrador to AK, south to n NJ, PA, and upland of NC, TN, west to IA and OR; also in Eurasia.

Distribution in New Jersey: A few stations in the northwestern part of the state.

Comments: This species was considered rare in New Jersey, although globally secure, by the New Jersey Natural Heritage Program (Snyder 1988).

In the field, the basal free pinnae point forward, adding another character to distinguish northern from southern beech fern. Northern beech fern, as the name implies, occurs in cool places in New Jersey, which is near the southeastern geographic limit of the species. The soil is probably acid, as suggested by Wherry (1961); however, some stations are in the limestone region.

T. phegopteris is an apogamous triploid, with $3x = 90$ chromosomes; this is the number reported for almost all localities from which chromosome studies have been made, but Mulligan and Cody (1982) reported apogamous tetraploids from Canada with $4x = 120$ chromosomes.

Thelypteris phegopteris (L.) Slosson

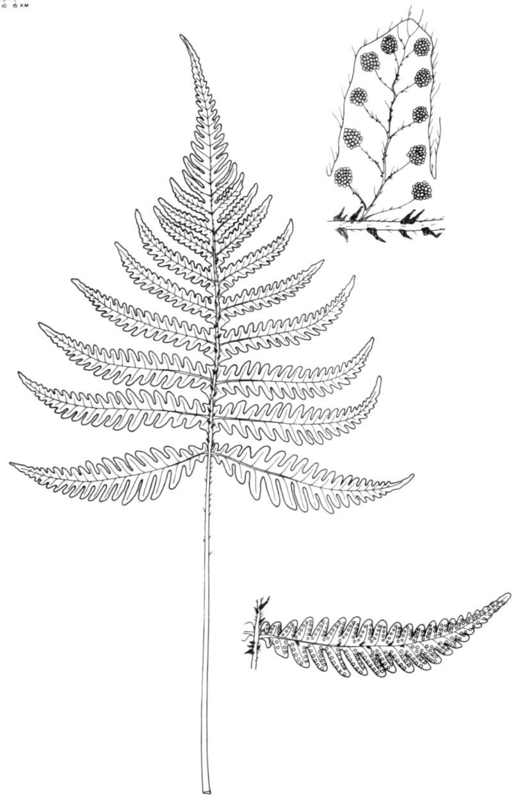

Thelypteris noveboracensis (L.) Nieuwl.

New York Fern

Description: Fronds narrowly elliptic, tapering to apex and base, with the lowest pinnae reduced to wings or auricles, pinnate-pinnatifid, up to 60 cm long; fertile fronds with pinnae flat, rounded at apex; indusia present.

Habitat: Mesic to damp woods, in full to partial shade.

Range: NF to MI south to GA, MS and AR.

Distribution in New Jersey: Common to abundant almost throughout the state; in all counties except Hudson, and lacking a recent record for Cape May.

Comments: This is one of the most common and, fortunately, most easily identified ferns in the state. The only other fern that tapers to the base is ostrich fern (*Matteuccia*), which is much larger, has dimorphic fronds, and is uncommon in the state.

The chromosome number of *T. noveboracensis* is $2n = 54$; thus this species is diploid.

Thelypteris noveboracensis (L.) Nieuwl.

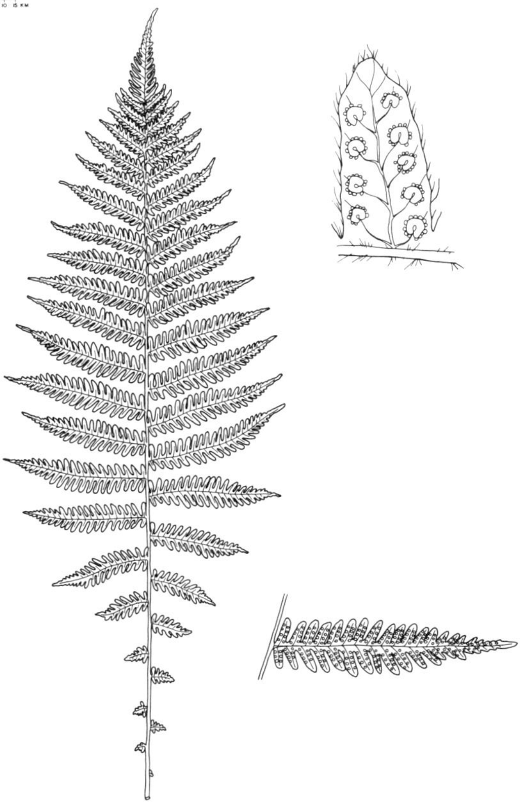

Thelypteris simulata (Davenp.) Nieuwl.

Massachusetts Fern

Description: Fronds lanceolate-ovate, truncate at the base, pinnate-pinnatifid, up to 60 cm long; fertile fronds with pinnae flat, rounded at apex; vegetative segments with veins simple (not forked); indusia present, glandular.

Habitat: Acid swampy woods and bogs, usually in full to partial shade.

Range: ME and NS, mostly along the coast to e VA, inland locally in NY, ne PA, n NJ, MD, and WI.

Distribution in New Jersey: Scattered in the northern counties; more common in the counties bordering the Delaware river and along the Atlantic Coast.

Comments: Massachusetts fern is found in acid swamps, and in New Jersey is often associated with Atlantic white cedar (*Chamaecyparis thyoides*), at the edges of—but infrequently *in*—the Pinelands, along the coast, or in the Ridge and Valley Province. Where marsh fern and Massachusetts fern grow together, the distinction in the fertile blades is readily apparent. Otherwise it is important to look for the simple or forked veins in the segments to determine the species.

The chromosome number of *T. simulata,* as reported by A. F. Tryon and R. Tryon (1973), is $2n = 128$; thus this species is tetraploid.

Thelypteris simulata (Davenp.) Nieuwl.

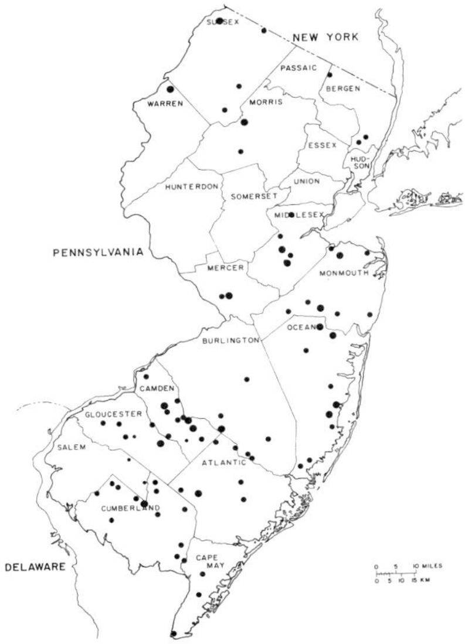

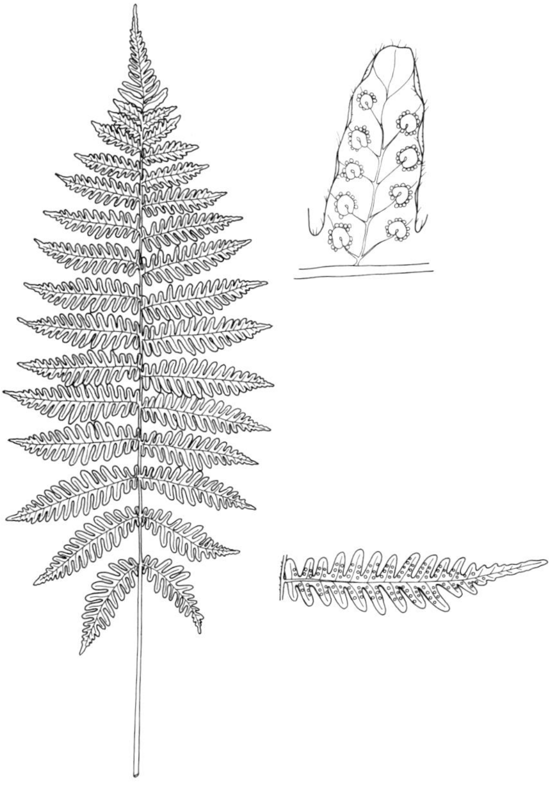

Thelypteris palustris Schott var. *pubescens* (Laws.) Fern.

Marsh Fern

Description: Fronds lanceolate-ovate, truncate at the base, pinnate-pinnatifid, up to 80 cm long; fertile fronds with segments often revolute and appearing acute at the apex; vegetative pinnae with veins in each segment once forked; indusia present, not glandular.

Habitat: Marshes, swamps, and damp thickets, in partial shade to full sun.

Range: NF to MB, south to FL, TX, and OK; other varieties in Eurasia, Africa, and New Zealand.

Distribution in New Jersey: Abundant throughout the state, with records for all counties (recent records lacking for four scattered counties).

Comments: Marsh fern often occurs in the freshwater edge of salt marshes along the coast. It is one of New Jersey's most sun-tolerant ferns, often growing in full sun in wet meadows. The North American variety differs only slightly from the European var. *palustris;* however, Holttum (1983) considered the North American plants to be a different species, *T. thelypterioides* (Michx.) Holub.

The chromosome number of *T. palustris* is $2n = 70$; thus this species is diploid.

Thelypteris palustris Schott
var. *pubescens* (Laws.) Fern.

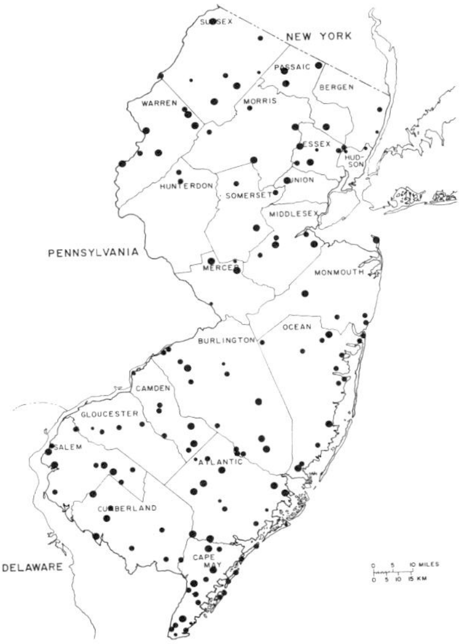

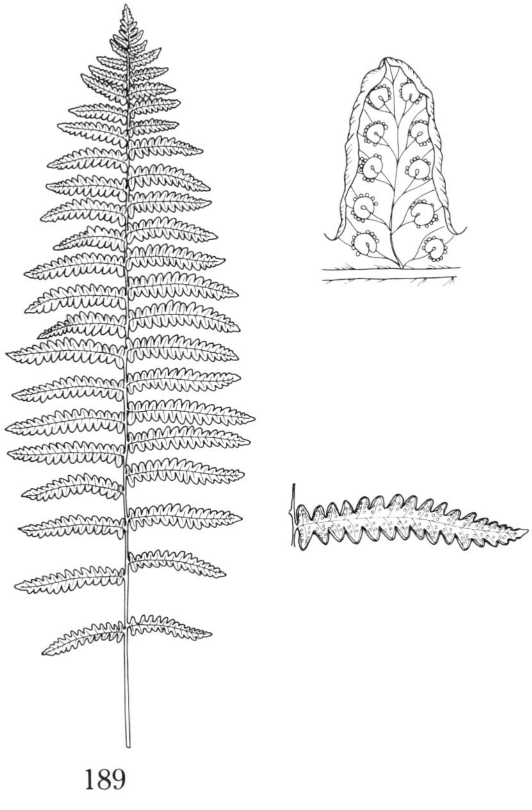

Asplenium

Small ferns, usually in rocky areas such as cliff faces or railroad cuts. Rhizome erect; fronds at least partly evergreen, sori elongate with lateral indusia. *Asplenium* is a very large genus, with most species in the tropics. New Jersey species include several widespread Northern Hemisphere species, as well as a distinctive complex called the Appalachian aspleniums. The classic study of this complex by Wagner (1954) is well known and requires no review. A key to the species and hybrids of the original "triangle" was presented by Wagner and Darling (1957).

Key to Species

1. Blades simple (not cut into separate segments) **2**
2. Blades entire except for basal lobes, rooting at prolonged tip .... *A. rhizophyllum*
2. Blades pinnatifid, tip not greatly prolonged, and usually not rooting . **3**
3. Blades regularly lobed, with lobes gradually decreasing toward leaf apex .... *A. pinnatifidum*
3. Blades irregularly lobed, with longer and shorter lobes erratically placed along the blade .... *A.* × *ebenoides*
1. Blades at least once pinnate (cut into separate pinnae) **4**
4. Blades once pinnate, pinnae entire or auriculate, blades linear lanceolate **5**
5. Pinnae oblong, at least 3 times as long as wide, auriculate........... .... *A. platyneuron*
5. Pinnae nearly round, only 1–2 times as long as wide, not auriculate or slightly so....................................... *A. trichomanes*
4. Blades pinnate-pinnatifid to tripinnate; blades oblong lanceolate to triangular **6**
6. Ultimate segments wedge-shaped, on stalks up to 3 mm long (on calcareous rocks) .... *A. ruta-muraria*
6. Ultimate segments ovate, short-stalked or sessile (on noncalcareous rocks).. **7**
7. Stipes and lower part of rachis dark; pinnae often with an enlarged basal lobe; blades oblong lanceolate........ *A. bradleyi*
7. Stipes dark at base, green above, rachis green; pinnae symmetrical; blades triangular.......................... *A. montanum*

Asplenium rhizophyllum L.

[*Camptosorus rhizophyllus* (L.) Link]
Walking Fern

Description: Fronds simple, 10–30 cm long, cordate and auriculate at base, prolonged into a slender apex, with a bud at the tip producing a new plant; veins areolate; sori elongate along the veins but appearing scattered on the lower frond surface.

Habitat: Boulders, ledges, and rock outcrops, most often of limestone, occasionally on red shale or traprock, usually in moist shaded locations.

Range: s PQ to MN, south to GA and OK.

Distribution in New Jersey: Most common in the limestone belt of Sussex and Warren counties; older records from traprock in central and northeastern New Jersey.

Comments: The name walking fern is appropriate, because the apical bud roots readily. Large colonies can be found nearly covering limestone boulders in rich moist woods of Sussex County.

The areolate veins and entire leaves have been used to segregate this species into the genus *Camptosorus,* with an additional species, *C. sibericus,* native to eastern Asia. The morphological distinctions are minor, and the frequency of hybrids between *A. rhizophyllum* and other Appalachian aspleniums support placement of this species in genus *Asplenium.*

The chromosome number of *A. rhizophyllum,* $2n = 72$, is diploid.

The hart's-tongue fern, *Phyllitis scolopendrium* (L.) Newm., sometimes placed in the genus *Asplenium,* is very rare in the New World, occurring in NY, ON, MI, e TN, and n AL. In 1936, members of the American Fern Society (Benedict 1936) planted this species in the Springdale area of Sussex County, in habitat somewhat like that in which it occurs near Syracuse, NY. The plants have survived but not spread.

Asplenium rhizophyllum L.

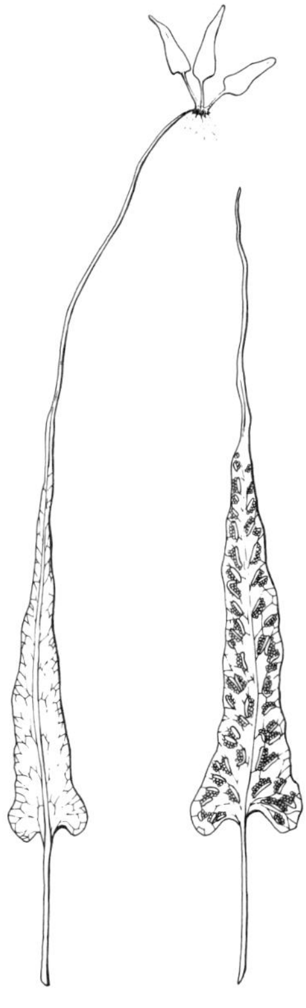

Asplenium pinnatifidum Nutt.

Lobed Spleenwort

Description: Fronds 7–20 cm long, lanceolate, tapering to the tip but not prolonged and rarely rooting, pinnatifid, with rounded lobes; stipe brown below, green above.

Habitat: Gneiss ledges and boulders.

Range: nw NJ and e PA, south to GA, OK, and MO.

Distribution in New Jersey: Rare; one recent record in Hunterdon County, and older records in Sussex County. The Sussex County station is the northeastern geographic limit of the species.

Comments: *A. pinnatifidum* was listed by Snyder and Vivian (1981) as "vulnerable and declining" in New Jersey, with one current station. It was considered as endangered by Montgomery (1989) and Fairbrothers and Hough (1973), and listed as critically imperiled in the state by the New Jersey Natural Heritage Program (Snyder 1988). The combination of a specific habitat requirement of acid cliffs and the fact that it is near the northern limit of its geographic range accounts for this rarity.

Lobed spleenwort was one of several ferns discovered in the state by J. L. Edwards; he collected it in Sussex County in 1939. The Hunterdon County station was discovered by Clausen and Kezer in 1941. Presumably plants once occurred near Blairstown, Warren County, since the hybrid with *A. montanum* (*A.* × *trudellii*) has been found there.

A. pinnatifidum originated as the result of hybridization between *A. montanum* and *A. rhizophyllum,* and so is tetraploid, with $2n = 144$.

Asplenium pinnatifidum Nutt.

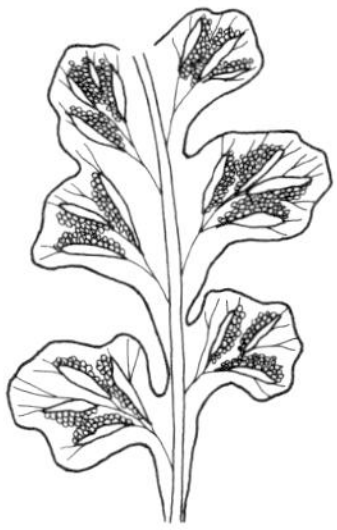

Asplenium platyneuron (L.) Oakes

Ebony Spleenwort

Description: Fronds lanceolate, once pinnate, somewhat narrowed at the base, narrowed to the tip; 20–50 cm long; usually with smaller vegetative evergreen fronds spreading at the base of the plant; stipe and rachis shiny dark brown; pinnae 0.5–3.0 cm long, sessile, oblong, auricled at base on upper side, serrate.

Habitat: Rocky or sandy open wooded slopes, cliffs, talus slopes, clearings and abandoned fields.

Range: ME to s ON, WI, IA, and KS, south to FL and e TX, also CO.

Distribution in New Jersey: Throughout the state, including recent records from all 21 counties. Common in the northern and southwestern counties. In the Pinelands nearly always associated with ruins and old cellar holes.

Comments: Unlike most of our spleenworts, ebony spleenwort is not restricted to rocky habitats, although it occurs on both basic and acid rock. Wooded slopes and roadside banks are frequent habitats in central and southern New Jersey. It can be abundant in old fields, along with *Botrychium dissectum* and goldenrod. In densely populated Hudson County, where ferns are very uncommon, it occurs on the only serpentine outcrop in the state.

A. platyneuron is diploid, with $2n = 72$.

Asplenium platyneuron (L.) Oakes

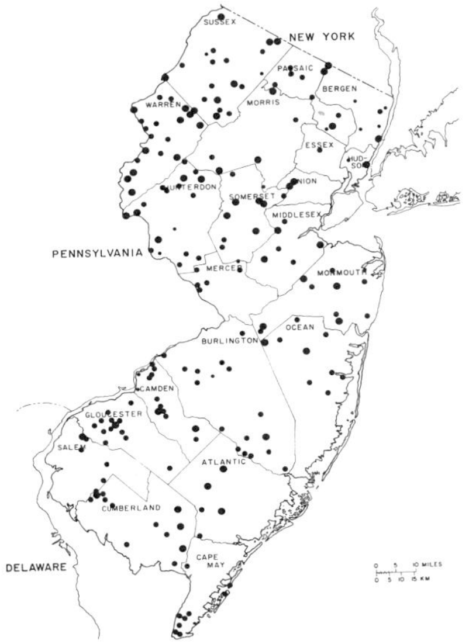

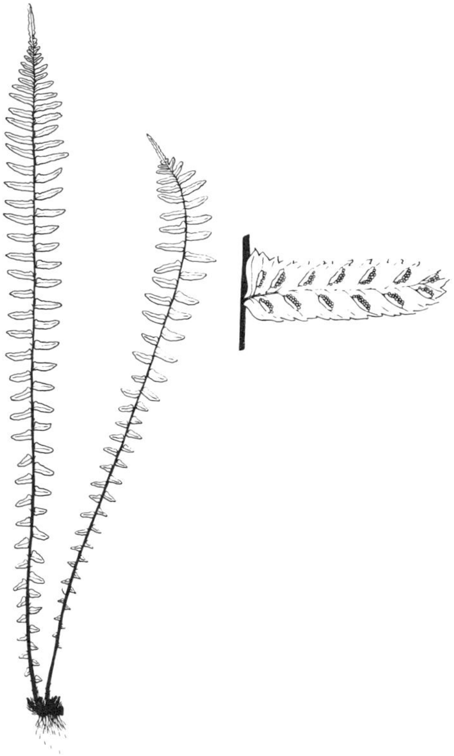

Asplenium trichomanes L.

Maidenhair Spleenwort

Description: Fronds linear, pinnate, slightly narrowed at base; 5–20 cm long; all alike and spreading; stipe and rachis wiry, shining dark purple-brown; pinnae small, ovate to nearly round, shallowly crenate on the margin.

Habitat: Ledges, crevices in cliff faces, and clefts of masonry; mostly calcareous to slightly acid substrate.

Range: PQ to MB, south to VA, GA, and AL, also in AR, OK, and the w United States; HI, Australia, New Zealand, and Eurasia.

Distribution in New Jersey: Frequent over the northern half of the state. Most of the sites in the Northeast were on traprock and have been destroyed by quarrying activity; still fairly common in the northwestern counties.

Comments: Recent studies in Europe and North America have revealed that there are two chromosome races of *A. trichomanes: A. trichomanes* subsp. *trichomanes* is diploid, with $2n = 72$, whereas *A. trichomanes* subsp. *quadrivalens* D. E. Meyer is tetraploid, with $2n = 144$. These are both fertile subspecies, and there are sterile triploid hybrids with 108 unpaired chromosomes. Unlike the other species of *Asplenium* that are of hybrid origin, these have originated by autopolyploidy, that is, doubling of the chromosomes within a single species. Although the diploid and tetraploid are very similar, they can usually be distinguished in the field, as described in the excellent report of Moran (1982). The tetraploid occurs mostly on calcareous rocks and has a somewhat thicker, blackish brown stipe, mostly opposite, oblong, and more crowded pinnae, and larger spores. The diploid occurs mostly on noncalcareous rocks and has a thinner coppery or bronze-red stipe, mostly alternate nearly round, more distant pinnae, and smaller spores. The triploid has been found mostly on limestone and is best distinguished by the aborted spores. R. C. Moran (personal communication) has found all three races in New Jersey: the tetraploid from Warren and Sussex counties; the diploid from Hunterdon, Warren, and Union counties (J. D. Montgomery has also collected it in Morris County); and the triploid from Warren County. The races are not distinguished on the distribution map; a careful study of the cytotypes in New Jersey would be a valuable contribution.

Asplenium trichomanes L.

Asplenium ruta-muraria L.

Wall-rue

Description: Fronds 5–10 cm long, ovate, bipinnate to tripinnate; stipe purple-brown at base, green above; segments stalked, wedge-shaped or fan-shaped, toothed at the tip.

Habitat: Ledges and cliffs of limestone.

Range: VT to n MI, south to n NJ, w VA, NC, AL, TN, and AR; also in Europe.

Distribution in New Jersey: Scattered on cliffs and ledges in the limestone region of Sussex and Warren counties.

Comments: This distinctive species usually occurs in drier sites than other aspleniums, often occurring with *Pellaea* (cliff-brake). The American plants are not very different from the European, although the American plants have been separated as var. *cryptolepis* (Fern.) Wherry. In Europe the plants are more common, often occurring on castle walls and in towns, hence the common name (Lat. *murus* = wall). Plants are also found in New Jersey on old railroad cuts in limestone areas.

A. ruta-muraria is tetraploid, with $2n = 144$.

Asplenium ruta-muraria L.

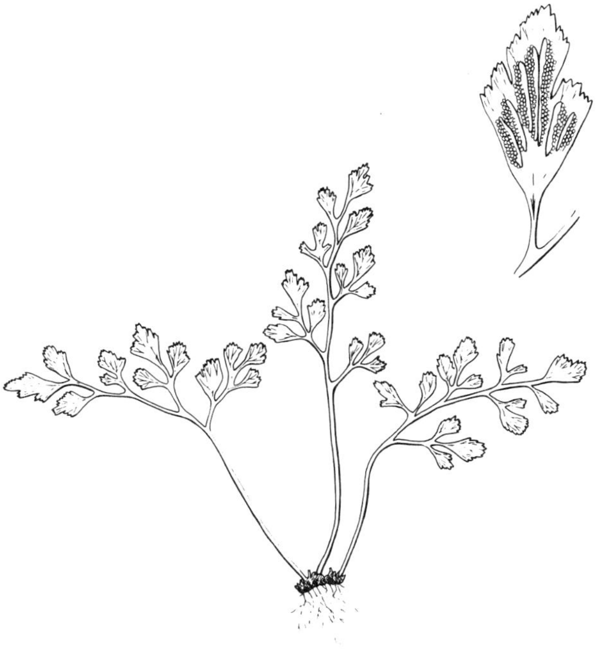

Asplenium bradleyi D.C. Eaton

Bradley's Spleenwort

Description: Fronds 5–20 cm long, oblong lanceolate, broadest above the base, pinnate-pinnatifid to bipinnate; stipe and lower rachis chestnut brown; pinnae nearly sessile, uneven at base, with a somewhat enlarged upper lobe or pinnule.

Habitat: Cliffs of sandstone or gneiss.

Range: se NY to s OH, south to GA, OK, and MO.

Distribution in New Jersey: Northwestern Morris County and northwestern Warren County; both sites extant.

Comments: The two recorded sites for the state were both discovered in 1935. The one in Warren County was found by members of the American Fern Society and reported by E. T. Wherry (1935). The Morris County site was discovered by J. L. Edwards.

A. bradleyi was listed as endangered in New Jersey by both Snyder and Vivian (1981) and Fairbrothers and Hough (1973); it was considered critically imperiled in the state and rare and local throughout its geographic range by the New Jersey Natural Heritage Program (Snyder 1988). Although there are records from the Shawangunk Mountains in southeastern New York, the New Jersey stations are near the northeastern geographic limit for this species.

A. bradleyi originated as a hybrid between *A. montanum* and *A. platyneuron* and is therefore tetraploid, with $2n = 144$. It backcrosses with *A. montanum* to produce the sterile hybrid *A.* × *wherryi*.

Asplenium bradleyi D. C. Eaton

Asplenium montanum Willd.

Mountain Spleenwort

Description: Fronds ovate-triangular, to 20 cm long, broadest at or near the base, usually bipinnate at least at base; stipes brown at base, green above, rachis green; pinnae with stalk up to 1 mm, symmetrical.

Habitat: Cliffs of noncalcareous rock.

Range: s CT and w MA to PA, WV, and e OH, south to GA and AL.

Distribution in New Jersey: Widely scattered in the northern part of the state, with recent records from Morris, Passaic, Union, and Warren counties.

Comments: The station on Mt. Tammany, above Delaware Water Gap, has been known since 1874. The Union County site was discovered by D. Snyder in 1979 and is notable because it is on traprock well east of the other localities in the state. New Jersey lies near the northeastern geographic limit for this species. This species was considered imperiled in the state by the New Jersey Natural Heritage Program (Snyder 1988).

A. montanum is diploid, with $2n = 72$.

Asplenium montanum Willd.

Asplenium × *ebenoides* R. R. Scott
A. platyneuron × *rhizophyllum*

[*Asplenosorus* × *ebenoides* (R. R. Scott) Wherry]
Scott's Spleenwort

Description: Fronds 6–25 cm long, irregularly lanceolate, attenuate at tip, pinnatifid (sometimes a few free pinnae at the base); stipes dark brown, the color usually continuing into the rachis; lobes variable in size and shape along the blade, rounded or acute, long or short; spores (in New Jersey) all abortive.

Habitat: Limestone boulders, ledges, and talus in woods, usually with both parents nearby.

Range: VT to NC, AL, and MO.

Distribution in New Jersey: Scattered localities in the limestone area of Warren and Sussex counties.

Comments: This is the classical fern hybrid, combining the characters of the parents and producing aborted spores. Individuals arise where *A. platyneuron* and *A. rhizophyllum* occur together but seem not to persist for long. The hybrid was first discovered by R. R. Scott along the Schuylkill River in southeastern Pennsylvania in 1864. The earliest collection from New Jersey was in 1879 by H. H. Rusby near Stillwater, Sussex County. E. T. Wherry placed the walking fern in *Camptosorus* and created the hybrid genus *Asplenosorus* for this hybrid. This usage has been expanded by Mickel (1979) to include other hybrids. Now that we better understand the evolution of the complex and the frequency with which hybrids occur, it is taxonomically appropriate to consider the complex of species within a single genus. New Jersey plants are the sterile diploid with 72 unpaired chromosomes; however, plants from Hale County, AL, are tetraploid and fertile.

Asplenium × *ebenoides* R. R. Scott

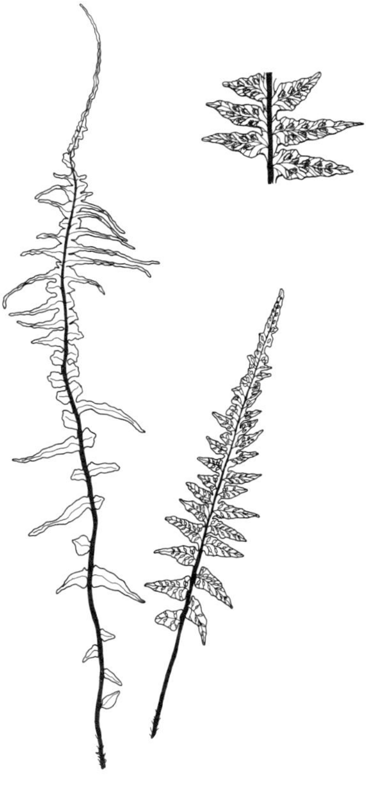

Asplenium × *trudellii* Wherry
A. montanum × *pinnatifidum*

[*Asplenosorus* × *trudellii* (Wherry) Mickel]

Description: Resembles *A. pinnatifidum,* with regularly lobed leaf blades and narrowly triangular shape; blades pinnate at base; stipes usually dark to the middle or slightly above; spores abortive.

Habitat: Sandstone cliffs.

Range: n NJ to OH, IL, KY, and AL.

Distribution in New Jersey: Once collected, by Macy Carhart in 1908 in western Warren County. Chrysler and Edwards (1947) report attempted recollection in 1935, without success, and plants have not been rediscovered to date.

Comment: As with most fern hybrids, aborted spores are the best characteristic by which to identify the plant as a hybrid.

Asplenium × *wherryi* D. M. Smith
A. bradleyi × *montanum*

Description: Difficult to distinguish from *A. bradleyi,* except for the aborted spores.

Habitat: Sandstone cliffs.

Range: nw NJ, se PA to KY.

Distribution in New Jersey: One station each in northwestern Morris County and western Warren County.

Comments: Although not collected in the state since 1940, both parents are still present at the Morris County site; presumably the hybrid could arise there again. W. H. Wagner, Jr., reported in 1956 that the vegetation at the Warren County station had changed and grapevines had overgrown the rocks where the fern originally grew.

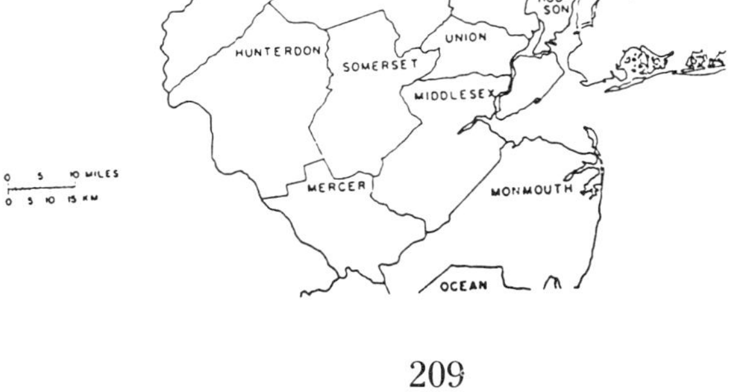

Matteuccia

Large woodland ferns. Rhizomes long-creeping, and branching to produce erect rhizomes at intervals that bear a crown of fronds; fronds dimorphic, the vegetative pinnate-pinnatifid, deciduous; fertile fronds pinnate-pinnatifid, persistent, with contracted segments surrounding the sori.

Matteuccia contains two species: one in eastern North America and Europe, and a second in eastern Asia.

Matteuccia struthiopteris (L.) Todaro var. *pensylvanica* (Willd.) Morton

[*M. pensylvanica* (Willd.) Raymond]
[*Pteretis nodulosa* (Michx.) Nieuwl.]
Ostrich Fern

Description: Fronds elliptic, narrowing to the apex and base, the lowest pinnae reduced to tiny flaps, vegetative fronds up to 1.5 m tall; stipes short, thick, nearly black; fertile fronds produced in late summer, shedding spores in late fall or early the following spring, narrower, but with similar shape.

Habitat: Alluvial soil in flood plain woods.

Range: NF to s AK, south to n VA, OH, MO, SD, and BC; also in Europe and e Asia.

Distribution in New Jersey: Scattered in Warren and Sussex counties; one or two sites recorded in five other counties.

Comments: Ostrich fern is one of New Jersey's most useful ferns. It is easily cultivated in a shady location, and, because of its ability to spread slowly, forms a bed of plants. The woody fertile fronds are frequently used in dry plant arrangements. This is also the only fern recommended for eating. The unrolling fiddleheads can be steamed like asparagus or eaten raw, but they become bitter as they reach 10–15 cm in length. Where plants are more abundant in New England and New Brunswick, Canada, they are canned and sold as a vegetable.

The chromosome number of *M. struthiopteris* is $2n = 78$.

Matteuccia struthiopteris (L.) Todaro
var. *pensylvanica* (Willd.) Morton

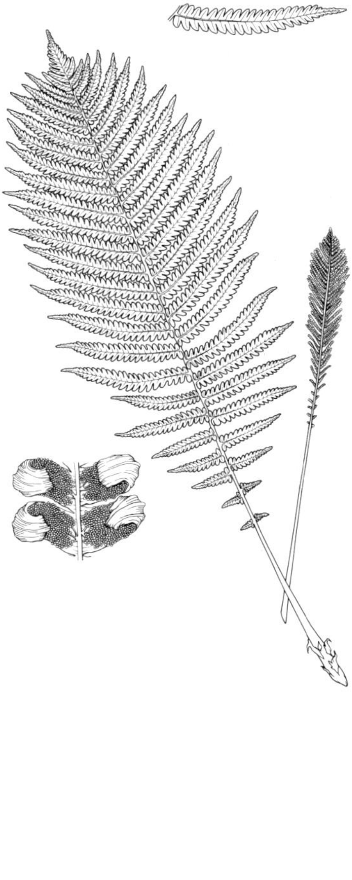

Onoclea

Medium-sized ferns of moist habitats. Rhizome long-creeping; fronds dimorphic, the vegetative fronds pinnatifid but not fully pinnate, deciduous; the fertile fronds pinnate-pinnatifid, with modified segments covering the sori. A single species in eastern North America and eastern Asia.

Onoclea sensibilis L.

Sensitive Fern

Description: Fronds dimorphic, the vegetative pinnatifid, deciduous, with the lobes wavy-margined; blade nearly triangular, up to 60 cm tall; veins areolate; fertile fronds produced late in the season and persistent all winter (spores shed in winter and early spring), up to 50 cm tall, tissue green at first, turning brown at maturity, the bead-like segments surrounding the sori.

Habitat: Usually in damp to wet places, including open marshes, meadows, and swamps, ditches, and openings in mesic forests; in full sun to partial shade, but fertile fronds found mostly in open sunny areas.

Range: NF to MB, south to FL and TX; also in e Asia.

Distribution in New Jersey: Abundant throughout the state; recent records from all 21 counties.

Comments: *Onoclea sensibilis* is perhaps our most abundant fern, growing in damp-to-wet areas in both shade (rarely fertile) to sunny open spaces. The common name refers to the sensitivity of the vegetative fronds to early frost. This species is also sensitive to pollution, producing morphological forms with part of the leaf fertile and part vegetative when sprayed or damaged.

The chromosome number of *O. sensibilis* is $2n = 74$.

Onoclea sensibilis L.

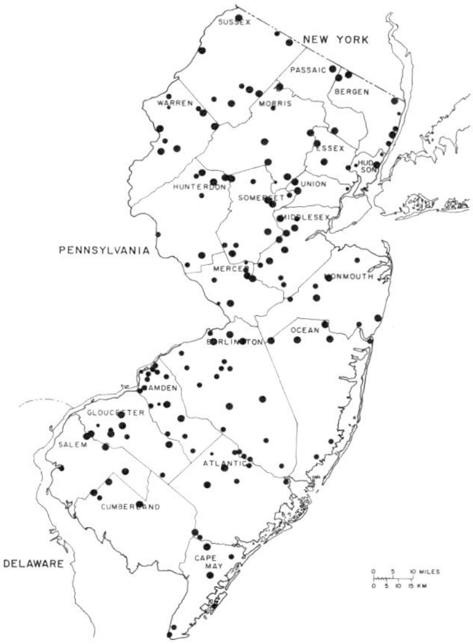

Athyrium

Ferns of moist woods, characterized by large size, deciduous fronds, stipes with two strap-shaped vascular bundles, and elongated sori covered by lateral indusia. The two vascular bundles at the base of the stipe (easily seen by snapping off a stipe at the base) are a useful diagnostic character to distinguish *Athyrium* from similar woodland ferns. Rhizomes are ascending or short-creeping in species from New Jersey.

The three species of *Athyrium* in New Jersey are not closely related. Kato (1977; see also Kato and Iwatsuki 1983) has suggested that they belong to three different genera, each of which has relatives in eastern Asia. Kato and Darnaedi (1988) concluded that *A. pycnocarpon* is a member of the largely tropical genus *Diplazium,* most closely related to a species in eastern Asia.

Key to Species and Varieties

1. Blades once-pinnate, pinnae undivided.................. *A. pycnocarpon*
1. Blades pinnate-pinnatifid to tripinnate.................................. **2**
  2. Blades pinnate-pinnatifid; sori straight .... *A. thelypterioides*
  2. Blades bipinnate to tripinnate; sori curved or hooked at the end .... *A. filix-femina*
    3. Blades broadest at or near the base (lowest pinnae the longest); spores black or dark brown....... *A. filix-femina* var. *asplenioides*
    3. Blades tapering at base (longest pinnae at about the middle of the blade); spores golden-yellow to light brown ....*A. filix-femina* var. *angustum*

Athyrium pycnocarpon (Spreng.) Tidestr.

Diplazium pycnocarpon (Spreng.) Broun
Glade Fern, Narrow-leaved Spleenwort

Description: Fronds to 1 m long, once-pinnate, pinnae entire; vegetative fronds spreading, later fertile fronds (when present) more erect and taller; sori elongate on both sides of the midrib; rhizome short-creeping.

Habitat: Rich damp woods; in New Jersey mostly on calcareous soil.

Range: sw PQ to NM, south to VA, GA, AL, LA, and e KS.

Distribution in New Jersey: Rare; occurring at half a dozen stations in the limestone area of Sussex County, with one older record for Warren County and one for Bergen County (West Englewood, 1915).

Comments: Glade fern looks similar to Christmas fern (*Polystichum acrostichoides*), but the pinnae lack the auricles and the fronds are not evergreen. Glade fern is not difficult to grow in a shaded garden, but it is not as attractive as Christmas fern. The fronds tend to become yellowed in late summer. Plants should certainly not be transplanted from the few remaining populations in New Jersey. This species was listed as rare in New Jersey by Snyder and Vivian (1981) and as endangered by Fairbrothers and Hough (1973); it was considered critically imperiled in the state, although globally secure, by the New Jersey Natural Heritage Program (Snyder 1988). Fertile fronds are rarely produced; we have observed the Sussex County plants for more than 20 years and have not seen a fertile frond. Plants may reproduce slowly by rhizome division.

A. pycnocarpon has a chromosome number of $2n = 80$.

Athyrium pycnocarpon (Spreng.) Tidestr.

Athyrium thelypterioides (Michx.) Desv.

Silvery Spleenwort, Silvery Glade Fern

Description: Blade pinnate-pinnatifid, up to approximately 1 m long; sori elongate, usually straight, with matching lateral indusia; rhizome short-creeping (the plants thus forming colonies).

Habitat: Wooded slopes especially on stream banks, in alluvial soil.

Range: NS to MI and MN, south to GA, AL, and e MO.

Distribution in New Jersey: Fairly common in Sussex and Warren counties; less common but reported for most other counties in the northern half of the state.

Comments: The common name comes from the silvery appearance of the young indusia. Like other athyriums, silvery spleenwort produces a crown of vegetative fronds in the spring, followed by fertile fronds in midsummer. The fertile fronds tend to be slightly narrower and more erect than the earlier vegetative fronds. Among New Jersey's larger woodland ferns the sori are distinctive; vegetative blades can be distinguished by their regular degree of cutting and the two stipe bundles. Blades tend to be a duller, softer green than the lady ferns.

The chromosome number of *A. thelypterioides* is $2n = 80$.

This is one of the New Jersey ferns typical of rich, moist woods, especially along streams. The underlying rock type seems not to matter as much as depth and moisture of the soil, although soils on limestone are usually rich in some of the valleys of northern New Jersey.

Athyrium thelypterioides (Michx.) Desv.

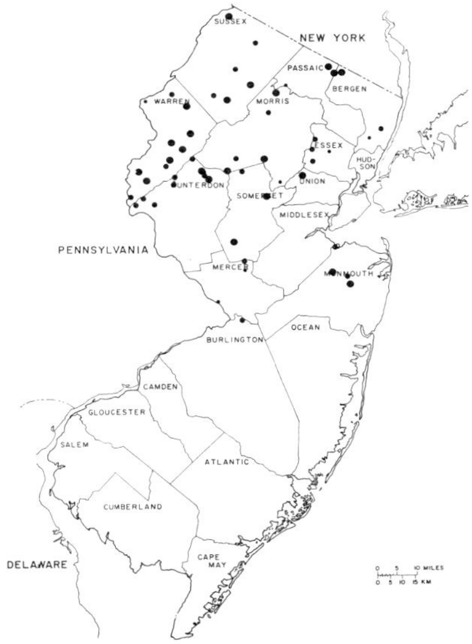

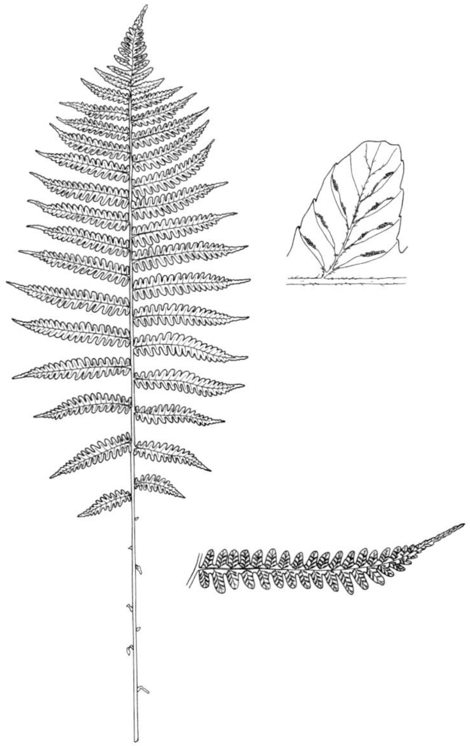

Athyrium filix-femina (L.) Roth var. *angustum* (Willd.) Moore

[*A. angustum* (Willd.) Presl]
[*A. filix-femina* var. *michauxii* (Spreng.) Farw.]
Northern Lady Fern

Description: Large woodland ferns; rhizomes ascending fronds up to 1 m long, bipinnate to tripinnate, with ultimate segments lobed; blades elliptic, with longest pinnae near the middle and tapering to base and apex; sori elongate and curved or hooked at the end; spores golden yellow to light brown.

Habitat: Moist woods of almost any composition; also on stone walls and in wet meadows.

Range: PQ to MN, south to s NJ, VA, and MO, becoming less common and replaced by var. *asplenioides* to the south.

Distribution in New Jersey: Common in most parts of the state except the Outer Coastal Plain and the Pinelands; recorded for all counties (only older records in Hudson and Cape May).

Comments: Northern and southern lady fern are difficult to distinguish at times; see the following discussion of southern lady fern. The type variety is native to Europe, and another very similar variety occurs in the western United States.

The chromosome number of *A. filix-femina* var. *angustum* is $2n = 80$.

The red-stemmed form, designated forma *rubellum,* is found as scattered individuals in most populations, thus is not taxonomically significant. There are a number of other forms cultivated in gardens, some being quite bizarre.

The name lady fern for these plants refers to their delicate appearance, in contrast to the male fern, *Dryopteris filix-mas*. (Note that *filix* means fern, whereas *felix* refers to cat!)

Athyrium filix-femina (L.) Roth
var. *angustum* (Willd.) Moore

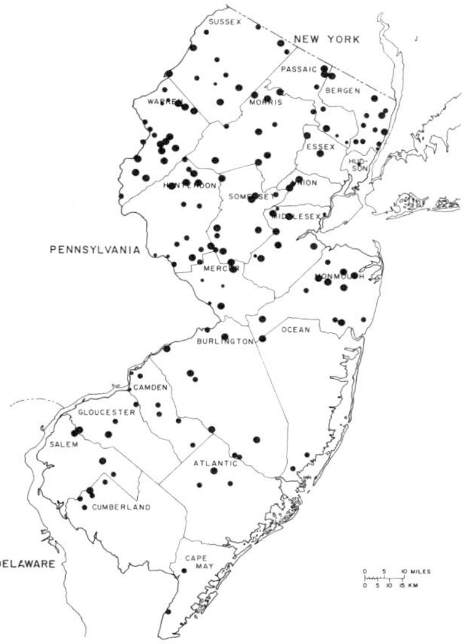

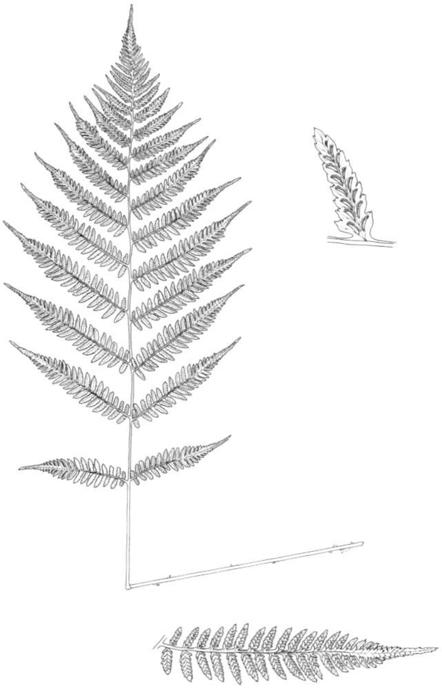

Athyrium filix-femina (L.) Roth var. *asplenioides* (Michx.) Farw.

[*A. asplenioides* (Michx.) A. A. Eaton]
Southern Lady Fern

Description: Large woodland ferns; rhizomes ascending blades up to 1 m long, bipinnate to tripinnate, with ultimate segments lobed; blades triangular, with longest pinnae at the base; sori elongate and curved; spores dark brown to black.

Habitat: Alluvial woods and moist wooded slopes along streams.

Range: FL to e TX, north to MA, IN, and MO; commoner southward.

Distribution in New Jersey: Mostly in the counties bordering the Delaware River on the Inner Coastal Plain.

Comments: In the southwestern portion of the state, the typical northern variety of lady fern grades into the southern variety. A few specimens are fairly typical of the southern variety, with basal pinnae as long as the middle ones, dark spores, and glandular ciliate indusia. Many specimens, however, combine the characters of the northern and southern varieties in various ways; such specimens cannot be "placed." These intermediates provide the evidence that these taxa are two varieties of a single species, rather than two species, as described by Wherry (1961). There is no evidence of hybridization, since all individuals produce good spores. Liew (1972) reported that the eastern American lady ferns were closely related and treated them as subspecies.

The chromosome number of *A. filix-femina* var. *asplenioides* is $2n = 80$.

Athyrium filix-femina (L.) Roth
var. *asplenioides* (Michx.) Farw.

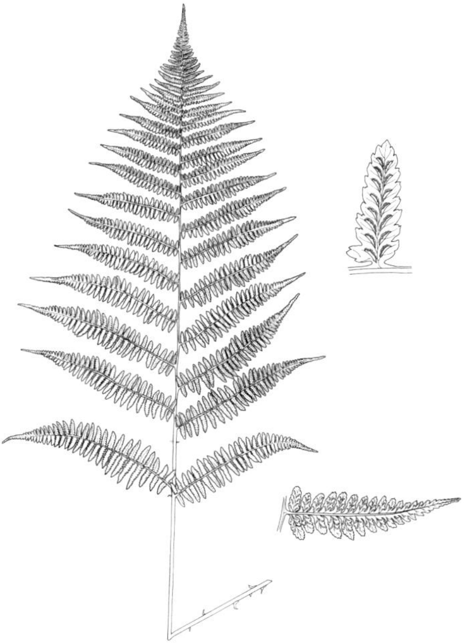

Gymnocarpium

Small woodland ferns. Rhizomes long-creeping, blades triangular, bipinnate-pinnatifid at base, pinnate-pinnatifid above, deciduous; sori round, dorsal, without indusia.

Gymnocarpium is a small genus of the North Temperate Zone.

Gymnocarpium dryopteris (L.) Newm. var. *dryopteris*

[*Phegopteris dryopteris* (L.) Fee]
[*Thelypteris dryopteris* (L.) Slosson]
[*Dryopteris disjuncta* (Ledeb.) Morton]
Oak Fern

Description: Fronds triangular, the lowest pair of pinnae much larger than those above, so that the blade appears three-parted, up to 35 cm tall; blades bipinnate-pinnatifid (basal pinnae) to pinnate-pinnatifid (above).

Habitat: Cool rocky woods and ravines in acid soil.

Range: Greenland and n Labrador, to AK, south to n NJ, upland VA, n OH, MI, WI, SD, NM, AZ, and OR; also in Eurasia.

Distribution in New Jersey: Uncommon in Sussex and Warren counties, and a single record for Hunterdon, Morris, and Passaic counties; an old specimen from Burlington County from an old well.

Comments: This species has been placed in *Dryopteris* because of the similar sorus shape (Fernald 1950), and in *Thelypteris* because of the sorus and blade shape similar to *T. phegopteris*. However, this species is related to *Athyrium* and *Cystopteris* and not to either *Dryopteris* or *Thelypteris* (Lellinger 1985; Crabbe et al. 1975).

The chromosome number of *G. dryopteris* is $2n = 160$.

Material from New Jersey is var. *dryopteris;* var. *disjunctum* (Rupr.) Ching occurs in the northwestern United States.

Gymnocarpium dryopteris (L.) Newm.

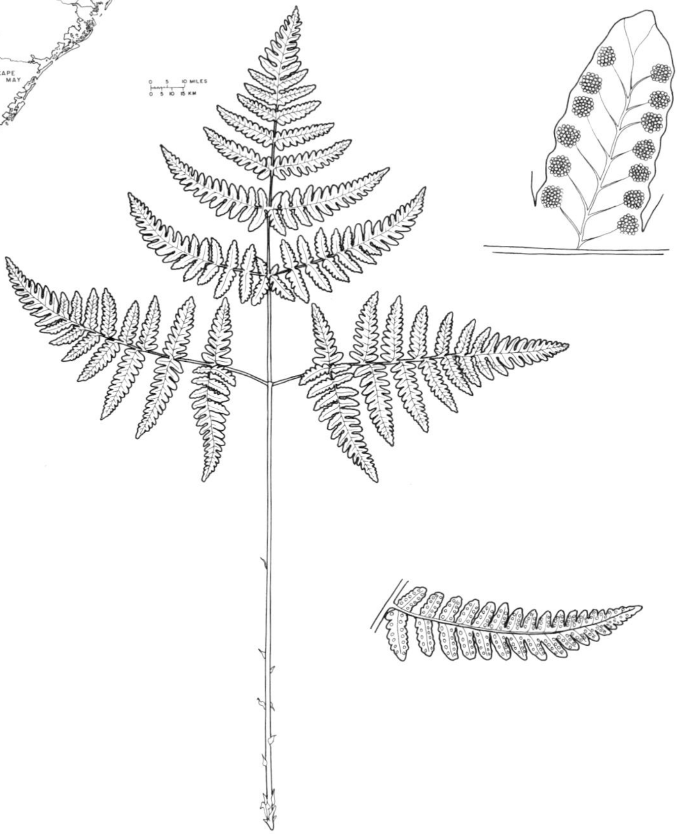

Cystopteris

Small to medium-sized rock or woodland ferns. Rhizomes short- to long-creeping, blades bi- to tripinnate, deciduous; sori round, with a scale-like indusium attached at the side of the sorus.

Cystopteris is a small genus, but contains a polyploid complex that is presently not fully understood. The three species in New Jersey are distinct morphologically and ecologically; however, their relationships with other species in the complex remain uncertain.

Key to Species

1. Blades linear-lanceolate, long-tapering at the tip, bearing bulblets along the rachis near the tip .... *C. bulbifera*
1. Blades ovate-lanceolate, short-tapering at the tip; bulblets lacking...... **2**
  2. Rhizomes short-creeping, not extending beyond stipes of the current season; plants of cliffs, ledges, or rocky woods .... *C. tenuis*
  2. Rhizomes long-creeping, extending (in summer and autumn) 1–4 cm beyond the stipes of the current season; plants of alluvial woods .... *C. protrusa*

Cystopteris bulbifera (L.) Bernh.

Bulblet Fern

Description: Rhizomes short-creeping; fronds bipinnate-pinnatifid, up to 60 cm long; stipes dark purple at the base, usually lighter above; blades linear-lanceolate, long-tapering at the apex, bearing one to several green bulblets 2–5 mm in diameter borne along the rachis near the apex on the underside of the blade.

Habitat: Shaded cliffs and talus of calcareous rock.

Range: NF to MB, south to n NJ, PA, and mountains of GA, TN, and AR; TX to AZ.

Distribution in New Jersey: Common in the limestone region of Sussex and Warren counties.

Comments: Most ferns have a mechanism for vegetative propagation by division of the rhizome. This occurs in bulblet fern; however, the plant also produces "bulblets," which are essentially leaf buds that

Cystopteris bulbifera (L.) Bernh.

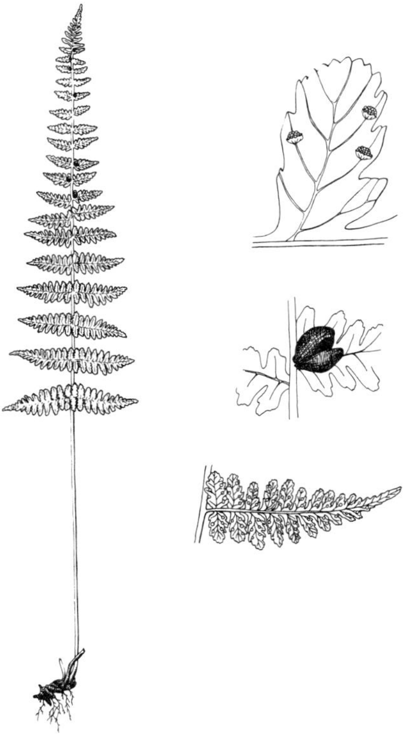

break off to produce new individuals (if they land in a favorable environment). This species can be propagated by these bulblets, but the plants can become aggressive in the garden, and the fronds become "tired" and thus not attractive in the late summer.

The chromosome number of *C. bulbifera* is $2n = 84$.

Cystopteris tenuis (Michx.) Desv.

[*C. fragilis* (L.) Bernh. var. *mackayi* Laws.]
Mackay's Fragile Fern

Description: Rhizomes short-creeping, the fronds of the current season clustered within 1 cm of the rhizome tip; fronds bipinnate-pinnatifid, up to 40 cm long; stipes dark brown at the base, brown or straw-colored above the base; blades lanceolate, acute, but not long-tapering at the tip, lacking bulblets.

Habitat: Crevices of cliffs, ledges, rocky woods, and soil banks on wooded slopes; substrate calcareous to acid.

Range: NF to AK, south to GA, AL, OK, AZ, and CA.

Distribution in New Jersey: Common in the northwestern part of the state, scattered on the Piedmont and Inner Coastal Plain; only one record on the Outer Coastal Plain (from masonry of an old well).

Comments: This species is a member of the "*Cystopteris fragilis* complex" (Lellinger 1985), which is currently being investigated (Haufler 1985; Haufler and Windham 1991). *C. tenuis* is tetraploid, with $2n = 84$; *C. protrusa* is one of the parents, but the other is unknown. As in other polyploid species complexes, it is reasonable to treat each taxon as a species, since each is distinct cytologically, morphologically, and (in *Cystopteris*) chemically.

Cystopteris tenuis (Michx.) Desv.

Cystopteris protrusa (Weath.) Blasdell

[*C. fragilis* (L.) Bernh. var. *protrusa* Weath.]
Lowland Fragile Fern

Description: Rhizomes long-creeping, with the growing tip of the rhizome extending 1–4 cm beyond the cluster of fronds of the season by midsummer; fronds bipinnate-pinnatifid, up to 40 cm long; stipes dark at the base, usually straw-colored above; blades lanceolate, acute, but not long-attenuate at the apex, without bulblets.

Habitat: Alluvial soil in flood plain woods.

Range: NJ to WI, south to VA, AL, LA, and NE.

Distribution in New Jersey: Rare; scattered localities in northwestern New Jersey along the Delaware River, and in Sussex and Warren counties.

Comments: *C. protrusa* was listed as of uncertain status by Snyder and Vivian (1981), but it is known from several localities in the state (Montgomery 1989); it was considered imperiled in the state by the New Jersey Natural Heritage Program (Snyder 1988). Protrusa refers to the rhizome protruding (extending) beyond the cluster of fronds.

This species hybridized with *C. bulbifera* to form *C.* × *tennesseensis* Shaver. This hybrid, known mostly from the southeastern United States, occurs as far north as central and eastern Pennsylvania. It could occur in New Jersey, since both parents are present in the state. The hybrid has abortive bulblets.

The chromosome number of *C. protrusa* is $2n = 84$, and it is one of the parents of *C. tenuis* (see our Comments on the latter species in the preceding discussion).

Cystopteris protrusa (Weath.) Blasdell

Woodsia

Small ferns of rocky habitats. Rhizomes short-creeping; fronds deciduous, pinnate-pinnatifid to bipinnate-pinnatifid; sori dorsal, round, with a cup-like indusium (often disappearing as the sori mature). Two species occur in New Jersey.

Key to Species

1. Blades nonglandular, with jointed hairs and scales on the lower surface; stipes with a joint, breaking off to form stubble of uniform length .... *W. ilvensis*

1. Blades glandular, lacking jointed hairs or scales on the lower surface; stipes lacking a joint, breaking off to form stubble of uneven length .... *W. obtusa*

Woodsia ilvensis (L.) R. Br.

Rusty Woodsia

Description: Rhizomes short-creeping; fronds up to 30 cm long, bipinnate-pinnatifid, at least in larger pinnae; stipes dark brown at base, straw-colored above, jointed (appearing as a minute swelling) above the base, so that stubble from old stipes is uniform in length; blades lanceolate, broadest above the base but below the middle, with jointed hairs and scales on the lower surface.

Habitat: Cliffs and rocky ledges in exposed or partially shaded places; not on limestone.

Range: Arctic Canada and AK, south to n NJ, PA, mountains of NC, west to MI, n IA, AB, and BC; also in northern Europe and Asia.

Distribution in New Jersey: Scattered in the northern third of the state.

Comments: The specific epithet, *ilvensis,* refers to the island of Elba in the Mediterranean, noted as the place where Napoleon Bonaparte lived after his abdication (May 1814–March 1815), and from where this species was originally described.

This species is not as common in New Jersey as the blunt-lobed

Woodsia ilvensis (L.) R. Br.

woodsia. One of the earliest fern records from New Jersey specific as to locality is a specimen of this species at the Academy of Natural Sciences of Philadelphia from the "Falls of the Passaic" (Paterson), collected by B. S. Barton in 1806.

The chromosome number of *W. ilvensis* is $2n = 82$.

Woodsia obtusa (Spreng.) Torr.

Blunt-lobed Woodsia

Description: Rhizomes short-creeping; fronds to 35 cm long, usually bipinnate-pinnatifid, at least at the base; stipes straw-colored above the base, not jointed, so that the stipes break off at irregular lengths; blades ovate to elliptic, broadest near the middle, glandular, but not scaly on the lower surface.

Habitat: Ledges and banks in woods or exposed places, on various types of rock as well as masonry.

Range: NB, sw PQ, MI, MN, and NE, south to GA and TX.

Distribution in New Jersey: Common in the northwestern part of the state, scattered on the Piedmont and Inner Coastal Plain; absent from the Outer Coastal Plain.

Comments: *W. obtusa* and *Cystopteris tenuis* are often confused in the field. They occupy similar habitats and are similar-sized ferns. The technical difference in the indusium (scale-like in *Cystopteris* and cup-like in *Woodsia*) is difficult to observe, especially since the indusia often disappear as the sori mature. The best distinctions are the following: the blades are broadest near the middle in *W. obtusa* but broadest near the base in *C. tenuis;* the pinnae are blunt at the tip in *W. obtusa,* as the name implies, and more acute in *C. tenuis;* and the stipe is more scaly in *W. obtusa* and more glabrous except at the very base in *C. tenuis.*

The chromosome number of *W. obtusa* is $2n = 156$.

Woodsia obtusa (Spreng.) Torr.

Polystichum

Medium-sized to large woodland ferns. Rhizomes ascending; fronds evergreen, pinnate to tripinnate, with ciliate to spiny-tipped teeth on the margins; stipes scaly; sori round, dorsal, with peltate indusia.

Polystichum is a medium-sized genus, with several species in the western United States, a few in Europe and at high elevations in tropical America, and many in Asia.

Polystichum acrostichoides (Michx.) Schott

Christmas Fern

Description: Fronds evergreen, lanceolate, up to 70 cm tall, pinnate; pinnae with a basal lobe on the upper side, sharply serrate; fertile pinnae only in the upper half of the blade, much smaller than the vegetative pinnae below; sori close together (appearing to cover the surface of the fertile pinnae).

Habitat: Rocky wooded slopes and rich moist or loamy woods, roadsides and stream banks.

Range: NS and PE, west to MI, WI, IA, and e KS, south to s FL and e TX.

Distribution in New Jersey: Abundant throughout most of the state (records for all counties except Hudson), but less common on the Outer Coastal Plain.

Comments: Christmas fern is one of the most familiar ferns in New Jersey. It is easy to cultivate, and the distinctive evergreen leaves make it a desirable plant for a woodland garden.

The chromosome number of *P. acrostichoides* is $2n = 82$.

Polystichum braunii (Spenner) Fee is found in Wayne County, PA, but has not been reported for New Jersey. The fronds are bipinnate; this species is also an attractive fern for a cool woodland garden.

Polystichum acrostichoides (Michx.) Schott

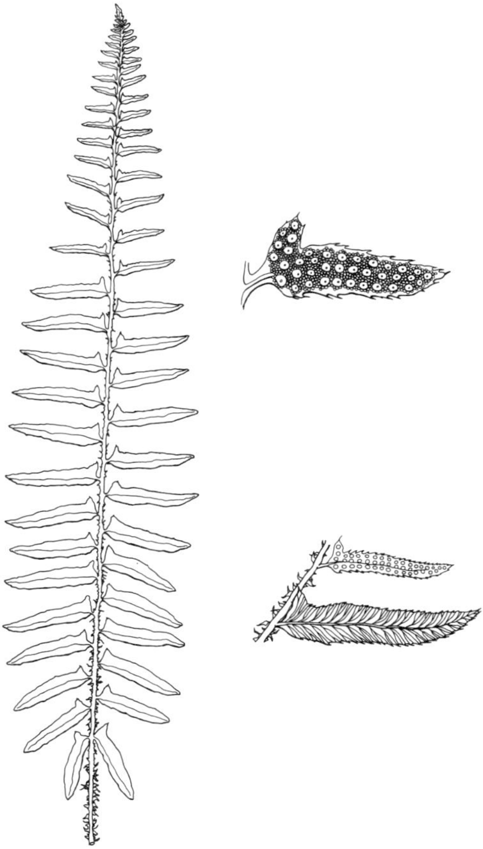

Dryopteris

Large woodland or swamp ferns. Rhizomes short-creeping or erect; fronds evergreen or deciduous, pinnate-pinnatifid to tripinnate; sori round, with orbicular reniform indusia.

Dryopteris is a large genus of temperate ferns (a few in the tropics at higher elevations). This is the most taxonomically complex genus in the northeastern United States, with 11 species and 27 hybrids in the region (Montgomery and Paulton 1981; Montgomery 1982a); 7 species and 19 hybrids have been found in New Jersey (Montgomery 1976). Until recently the species were poorly delimited, with some considered as varieties (Chrysler and Edwards 1947; Fernald 1950). Studies by Walker (1955, 1961, 1962), D. M. Britton (1976), and especially Wagner and his coworkers (Wagner 1971; Wagner and Wagner 1965; Wagner et al. 1969), have employed chromosome numbers and pairing patterns, as well as traditional morphological characteristics and population studies to establish the relationships among the species. The North American taxa are presently reasonably well understood taxonomically, although there remain a few unanswered questions (Montgomery and Paulton 1981). Diagrams of the relationships have been published by Wagner (1971, p. 166), Montgomery (in Montgomery and Paulton 1981), and Lellinger (1985, pp. 320–321). The base chromosome number in *Dryopteris* is 41. The species in New Jersey are diploid (41 pairs of chromosomes), tetraploid (82), or hexaploid (123), and hybrids have numbers ranging from 82 to 205, with single chromosomes or a mixture of pairs and single chromosomes.

The first problem in identification of a specimen of *Dryopteris* is to determine whether the plant is a fertile species or a sterile hybrid. Hybrids can be distinguished by their aborted spores and intermediate morphology; see the discussion of *Dryopteris* hybrids (below) for additional information. The key below, based on Montgomery (1982a), includes the seven species from New Jersey, as well as the "more common" hybrids (those with more than five records for the state). "Rare" hybrids (those with five or fewer records) can be identified using parental characters or the inclusive key by Montgomery (1982a). Descriptions, maps, and illustrations are provided for the eight "more common" hybrids; the eleven "rarer" hybrids are briefly described but not illustrated.

Key to Species and More Common Hybrids

1. Spores all or nearly all normal in size and shape (fertile species **2**
2. Blades pinnate-pinnatifid or bipinnate at base only **3**
3. Sori at or near the margin of the segments; blades evergreen, leathery in texture and blue-green in color .... *D. marginalis*
3. Sori halfway between the margin and midrib or closer to midrib of the segments; blades deciduous, or with smaller evergreen vegetative ones, herbaceous in texture and green in color **4**
4. Lowest pinnae ovate (broadest above the base); fertile fronds ovate to ovate-lanceolate, not with parallel sides; scales at base of stipe dark brown or with a dark brown stripe.............. **5**
5. Blade tapering abruptly to the tip, ovate; sori nearer the midvein than the margin of segments .... *D. goldiana*
5. Blades gradually narrowed to the tip, ovate-lanceolate; sori about midway between midvein and margin of the segments... *D. celsa*
4. Lowest pinnae triangular (broadest at base); fertile fronds lanceolate and parallel-sided; scales at base of stipe tan **6**
6. Basal pinnae $1\frac{1}{2}$–2 times longer than wide; pinnae of fertile fronds nearly in plane of blade .... *D. clintoniana*
6. Basal pinnae about as wide as long; pinnae of fertile fronds turned at an agle to plane of blade .... *D. cristata*
2. Blades bipinnate to tripinnate.. **7**
7. Indusia and axes of segments glandular; fronds evergreen; first basal pinnule of basal pinnae often shorter than second .... *D. intermedia*
7. Indusia and axes of segments not glandular; fronds deciduous; first basal pinnule of lowest pinnae often longer than second ....*D. carthusiana*
1. Spores aborted (hybrids).. **8**
8. Blades pinnate-pinnatifid, or bipinnate at base only **9**
9. Sori closer to margin than midrib **10**
10. Scales at base of stipe dark glossy brown with a pale border; basal pinnae lanceolate; blades broad, up to 30 cm wide .... *D. goldiana* × *marginalis* (*D.* × *neowherryi*)
10. Scales at base of stipe tan, dull, uniform in color; basal pinnae triangular; blades narrow, 10–20 cm wide.................... ....*D. cristata* × *marginalis* (*D.* × *slossonae*)
9. Sori closer to midrib than margin **11**
11. blades 2–3 times longer than wide; scales at base of stipe dark brown with paler broder .... *D. clintoniana* × *goldiana*
11. Blades more than 3 times longer than wide; scales at base of stipe tan to brown................... *D. clintoniana* × *cristata*
8. Blades bipinnate to tripinnate (bipinnate above the basal pair of pinnae) **12**

12. Fronds tripinnate . . . *D. carthusiana* × *intermedia* (*D.* × *triploidea*)
12. Fronds bipinnate-serrate to bipinnate-pinnatifid. **13**
13. Indusia and axes without glandular hairs . . . . *D. carthusiana* × *cristata* (*D.* × *uliginosa*)
13. Indusia and often axes with small glandular hairs **14**
14. Fronds small, fertile ones usually less than 20 cm wide, and with the longest pinnae 7–11 cm long . . . . *D. cristata* × *intermedia* (*D.* × *boottii*)
14. Fronds large, fertile ones usually more than 15 cm wide, and with the longest pinnae 10–15 cm long. . . . .*D. clintoniana* × *intermedia* (*D.* × *dowellii*)

Dryopteris marginalis (L.) Gray

Marginal Wood Fern

Description: Fronds pinnate-pinnatifid, sometimes bipinnate at the base, evergreen, blue-green in color and leathery in texture, up to 1 m long; blades ovate-lanceolate; stipe scales bright tan, in a dense tuft at the base; sori near the margin of the segments, indusia glabrous.

Habitat: Rocky wooded slopes and ravines, edges of woods, stream and road banks, and rock walls.

Range: e PQ to n MI, south to GA, AL, AR, OK,and KS.

Distribution in New Jersey: Abundant, especially in northern New Jersey. Recorded from all counties except Cape May; no recent collections from Hudson, Ocean, Atlantic, or Cumberland counties.

Comments: On rocky moist wooded slopes, this species is usually present. The evergreen fronds become flattened to the ground in winter, making conspicuous green patches on the otherwise brown slopes. Marginal wood fern makes an excellent garden or border plant, since the erect rhizome rarely divides and forms a large crown.

Large plants or late fronds (fronds produced after midsummer, often after injury) may have more finely divided blades, approaching bipinnate-pinnatifid. Such specimens can be confused with hybrids; however, the spores are not aborted, the sori are close to the segment margins, and less divided blades are also usually present on the same plant.

Dryopteris marginalis (L.) Gray

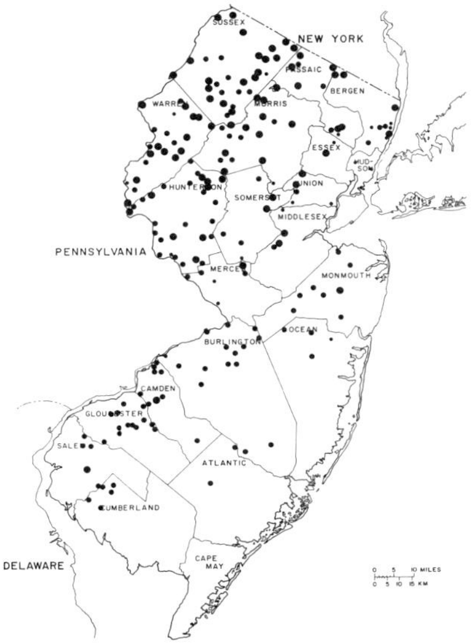

D. marginalis hybridizes readily: hybrids are recorded with all other species of the genus found in New Jersey. It is interesting that *D. marginalis* rarely hybridizes with *D. intermedia,* the species with which it most frequently occurs, but there is no satisfactory explanation.

D. marginalis is diploid ($2n = 82$).

Dryopteris goldiana (Hook.) Gray

Goldie's Wood Fern

Description: Fronds pinnate-pinnatifid, sometimes bipinnate at the base, deciduous, green and often mottled at the tip, up to 1.2 m long, ovate, abruptly tapering to the tip; pinnae ovate, widest above the base; scales glossy dark brown to almost black; sori nearer the midvein than the margin of the segments; indusia glabrous.

Habitat: Rich moist woods, especially in ravines, or at the edge of (but not in) swamps.

Range: w NB and sw PQ to se MN, south to s VA, NC, TN, KY, and IL.

Distribution in New Jersey: Uncommon; recent records only for Sussex and Morris counties.

Comments: Goldie's wood fern is one of New Jersey's largest and most attractive ferns. It is readily grown in a woodland garden but should not be dug from the few remaining wild populations. Chrysler and Edwards remarked in 1947 that it seemed strange to find that specimens of Goldie's wood fern were collected "near Newark" in 1862. This collection actually came from the "Orange Mountains" (Watchung Mountain), probably in West Orange.

D. goldiana is diploid ($2n = 82$) and is one parent of *D. celsa* and *D. clintoniana*. *D. goldiana* hybrids are rare in New Jersey because *D. goldiana* is uncommon; the hybrid with *D. marginalis* is often found when the parents grow together (Montgomery 1976).

Dryopteris goldiana (Hook.) Gray

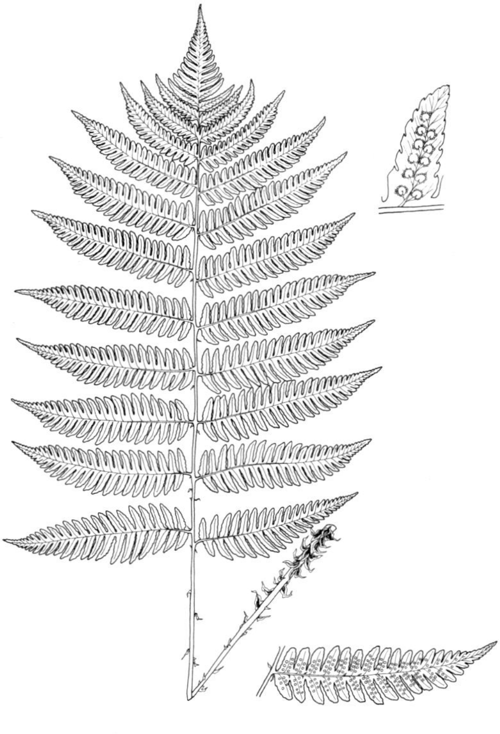

Dryopteris celsa (W. Palmer) Small

Log Fern

Description: Fronds pinnate-pinnatifid, deciduous, up to 1.2 m long, ovate, gradually tapering to the tip; pinnae ovate, widest above the base; scales dark brown or tan with a dark stripe; sori about midway between margin and midvein of the segments; indusia glabrous.

Habitat: Seepage slopes, hummocks, and logs in swamps.

Range: Mostly on the Piedmont or Coastal Plain, LA to SC, north to se VA, se PA, ne NJ, and disjunct in upstate NY and MI.

Distribution in New Jersey: Probably extirpated; three localities (one known from hybrids only) in Bergen County.

Comments: *D. celsa* was collected in New Jersey by Macy Carhart in a swamp north of Lodi in 1908 and by Philip Dowell in a swamp at West Englewood Station in 1909. During the next seven years numerous collections were made of both the species and of hybrids. There are no New Jersey collections after 1915, and the West Englewood locality was reported destroyed by 1948 (Dole 1948). In 1957 Robert Gaede sent living plants to W. H. Wagner, Jr., from an area being destroyed by construction of the Garden State Parkway; these were found to be hybrids involving *D. celsa* (Wagner 1971), although *D. celsa* was not collected. Although extensive searches were made in Bergen County (Montgomery 1975), the possibility exists that plants of log fern occur in wetlands in Bergen and Passaic counties. There is also a specimen in the Berlin–Dahlem Botanical Museum of *D. celsa* marked "Hudson Co., New Jersey," collected by M. Kuhn in 1868. This may represent another former population or an early collection from West Englewood Station, near the Bergen County–Hudson County line.

D. celsa is tetraploid ($2n = 164$).

Dryopteris celsa (W. Palmer) Small

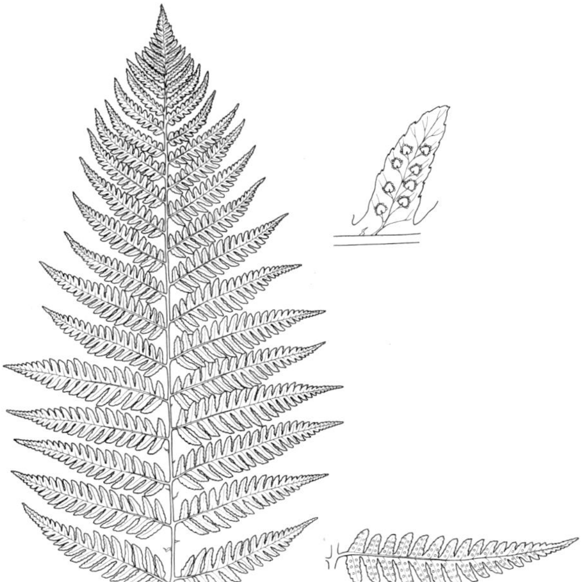

Dryopteris clintoniana (D.C. Eaton) Dowell

Clinton's Wood Fern

Description: Fronds pinnate-pinnatifid, up to 1 m long, lanceolate with nearly parallel sides; fertile fronds deciduous, with one to several smaller evergreen vegetative fronds; pinnae elongate-triangular, widest at base, about twice as long as wide, more or less in plane of blades, not twisted or only slightly so; scales light brown, sometimes with a dark brown center; sori about midway between margin and midvein of segments; indusia glabrous.

Habitat: Swampy woods, especially red maple swamps.

Range: Central ME to sw ON, MI, s WI, and MN, south to s New England, n NJ, PA, n OH, n IN; rare south of the Wisconsin Glacial Boundary.

Distribution in New Jersey: Uncommon; several locations in Sussex County, a few in Passaic, Morris, and Warren counties, and older records for Essex, Middlesex, and Monmouth counties.

Comments: Clinton's wood fern and crested wood fern often grow together in red maple swamps in the glaciated northwestern region of New Jersey. The larger size and more elongate lower pinnae distinguish *D. clintoniana,* but hybrid plants occur and blur the distinctions. Hybrids can be distinguished only by the aborted spores.

D. clintoniana is hexaploid ($2n = 246$); counts have been made from New Jersey (Montgomery 1975). The spores are larger than those of any other species of *Dryopteris* in the state. Spore size was used to distinguish specimens of *D. clintoniana* from *D. celsa,* where living material of the latter species was not available (Montgomery 1975).

Dryopteris clintoniana (D. C. Eaton) Dowell

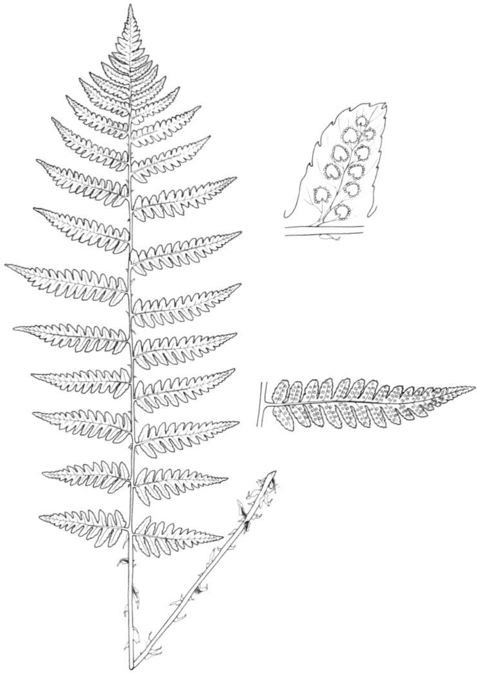

Dryopteris cristata (L.) Gray

Crested Wood Fern

Description: Fronds pinnate-pinnatifid, up to 70 cm long, narrow-lanceolate, with parallel sides; fertile fronds deciduous, with several smaller evergreen vegetative fronds forming a "rosette"; pinnae triangular, widest at base, about as wide as long, often twisted nearly horizontally in comparison to the erect fertile blades; scales light brown; sori about midway between the margin and midrib of the segments; indusia glabrous.

Habitat: In wooded swamps or open shrubby wetlands.

Range: NF to BC, south to NS, New England, Long Island, NY, VA, TN, west to IA, NE, nw MT, and n ID; circumpolar, in Europe, Siberia, and Japan.

Distribution in New Jersey: Common; records for all counties except Hudson and Somerset, but more abundant in the northern part of the state, and few recent records in southern New Jersey.

Comments: In the field, the tendancy for the pinnae to twist horizontally is very noticeable (and makes the plants difficult to photograph). *D. cristata* is more common than *D. clintoniana* and much more widespread in the state. Although the natural habitat is wet soil, plants grow readily in a moist shaded garden, where the effect of the twisted pinnae is striking.

D. cristata is tetraploid ($2n = 164$), and the chromosomes have been counted from New Jersey specimens (Montgomery 1976). One of the parents is *D. ludoviciana,* a southeastern species; the other is shared with *D. carthusiana* but has never been found. Wagner (1971) refers to it as "D. semicristata," since it constitutes half of *D. cristata.*

Dryopteris cristata (L.) Gray

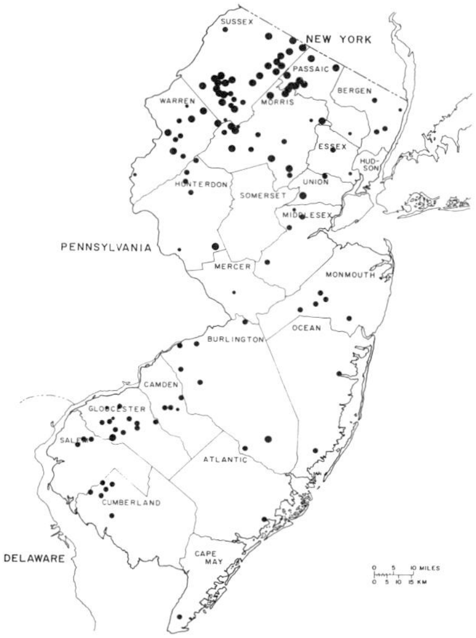

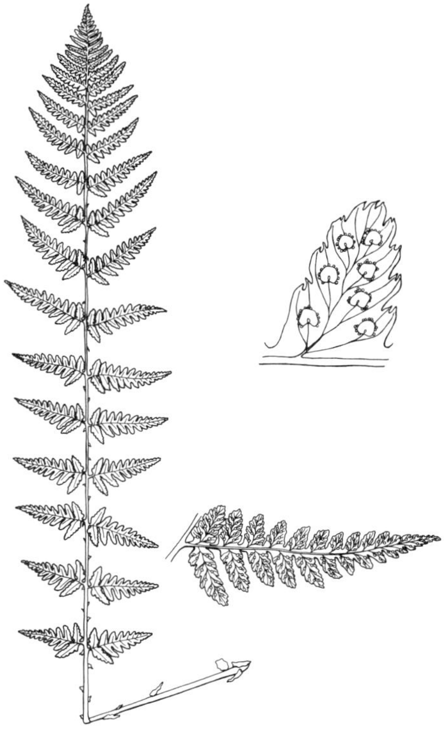

Dryopteris intermedia (Muhl.) Gray

[*D. spinulosa* (O. F. Muell.) Watt. var *intermedia* (Muhl.) Gray]
Evergreen Wood Fern, Fancy Fern

Description: Fronds tripinnate-pinnatifid, evergreen, up to 90 cm long, ovate; basal pinnae with first lower pinnule usually shorter than second; scales light brown; indusia and usually also the axes of the segments with minute glandular hairs.

Habitat: Moist rocky woodlands, especially in hemlock–hardwood ravines; also in swampy woods, but not in very wet places, and in conifer plantations.

Range: NF to w ON, and e MN, south to VA, AL, TN, west to IN, IL, and IA.

Distribution in New Jersey: Abundant, especially in northwestern New Jersey. Records from all counties except Middlesex, but recent records lacking from five additional counties.

Comments: *D. intermedia* and *D. carthusiana* have been confused, or included under the same species (as in Chrysler and Edwards 1947). Chromosome studies have shown that *D. intermedia* is diploid ($2n = 82$) and *D. carthusiana* is tetraploid; and, further, that *D. intermedia* is one of the parents of *D. carthusiana*. The best way to distinguish them in the field is to learn to recognize the glandular hairs ("glands") on the indusia and minor axes of *D. intermedia*. They look like very tiny hat pins.

Evergreen wood fern is an excellent native fern for the garden. It is abundant in the state, evergreen, and large enough to be showy. It spreads only very slowly, since the rhizome does not branch frequently.

Dryopteris intermedia (Muhl.) Gray

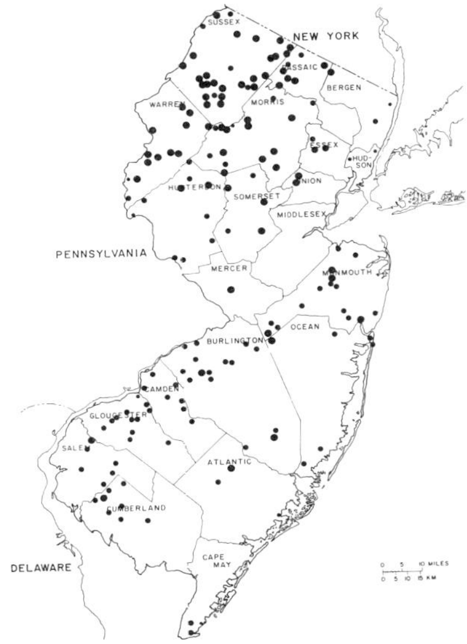

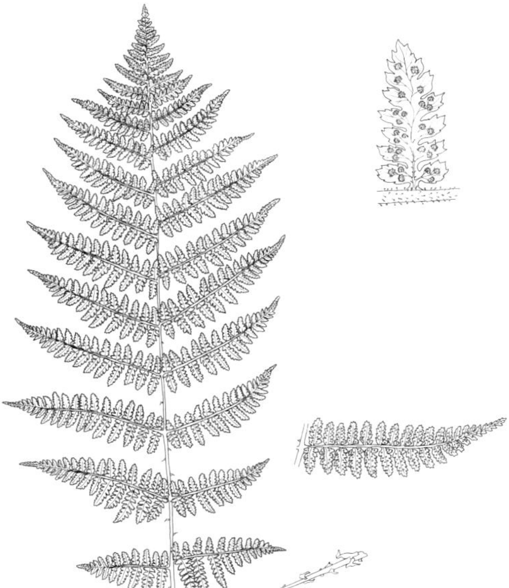

Dryopteris carthusiana (Vill.) H. P. Fuchs

[*D. spinulosa* (O. F. Muell.) Watt.]
[*D. austriaca* (Jacq.) Woynar]
Spinulose Wood Fern

Description: Fronds bipinnate to tripinnate-pinnatifid, deciduous, up to 70 cm long, ovate; basal pinnae with first lower pinnule often longer than second; scales light brown; indusia and axes glabrous, without glands.

Habitat: Swampy woods, especially red maple swamps; also on moist wooded slopes or stream banks; and not infrequently around old houses, foundations, and in conifer plantations.

Range: s Labrador to AB, south to NS, New England, VA, OH, IN, IL, MO, MT, and ID; also in Europe.

Distribution in New Jersey: Abundant; recorded for all 21 counties, but recent records lacking for Hudson and Cape May.

Comments: This species is often confused with *D. intermedia* (see the Comments in the foregoing discussion). *D. carthusiana* was called *D. spinulosa* by most authors until recently. Fraser-Jenkins (1980) has demonstrated that the name *D. carthusiana* applies to this taxon and, since this is the older name, it is the correct one.

D. carthusiana is tetraploid ($2n = 164$). *D. intermedia* is one parent, as indicated by chromosome pairing in the hybrid *D.* × *triploidea*. The other parent is shared with *D. cristata,* again as indicated by pairing in *D.* × *uliginosa* (see *D. cristata*).

Dryopteris campyloptera (Kunze) Clarkson has occasionally been reported from New Jersey, but Chrysler and Edwards (1947) reported that there are no verified herbarium records for the state. Populations are found as close as Wayne County in northeastern Pennsylvania, but at higher elevations (2500 ft) than occur in New Jersey.

Dryopteris carthusiana (Vill.) H. P. Fuchs

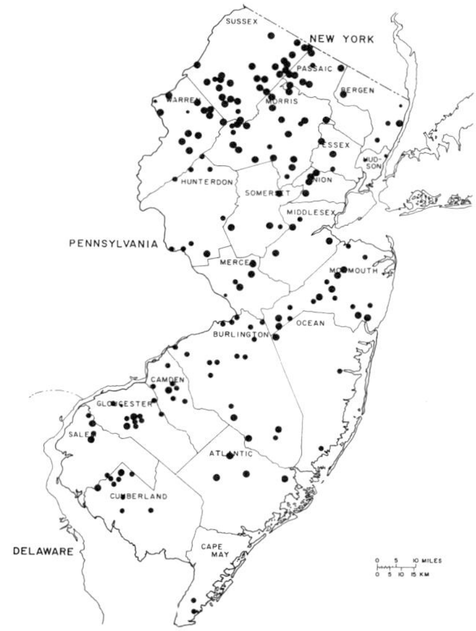

Dryopteris Hybrids

The *Dryopteris* hybrids known from New Jersey are arranged alphabetically by the parent taxa, except for *D. celsa* hybrids, which are presented together at the beginning. As explained earlier in the *Dryopteris* generic introduction, "more common" hybrids are fully treated whereas "rarer" hybrids are briefly noted as to occurrence in New Jersey.

Dryopteris celsa Hybrids

Five hybrids involving *D. celsa* are recorded from Bergen County: *D. celsa* × *cristata, D. celsa* × *goldiana, D. celsa* × *intermedia* [*D.* × *separabilis* (Palmer) Small], *D. celsa* × *marginalis* (*D.* × *leedsii* Wherry), and possibly *D. carthusiana* × *celsa* (Montgomery 1975, 1976). There are specimens of all taxa from Lodi, and all except *D. carthusiana* × *celsa* from West Englewood, collected in 1909–1915. The Gaede collection from Ridgewood includes only *D. celsa* × *cristata* and *D. celsa* × *goldiana* (Wagner 1971). See the previous Comments under *D. celsa* regarding the status of the species and its hybrids in New Jersey.

Dryopteris carthusiana × *clintoniana* *D.* × *benedictii* Wherry

This hybrid resembles the more abundant *D. clintoniana* × *intermedia,* both in habitat and morphology, but lacks the glandular hairs on the indusia and axes. There are five records from swamps in Sussex, Passaic, and Morris counties.

Dryopteris carthusiana × *cristata* *D.* × *uliginosa* (A. Br.) Druce

Description: Fronds deciduous, bipinnate at least at the base; basal pinnae triangular; glands absent on axes and indusia.

Habitat: Usually in wooded swamps or bogs.

Range: ME to VA and WV, west to MN and ND; also in Europe.

Dryopteris carthusiana × cristata

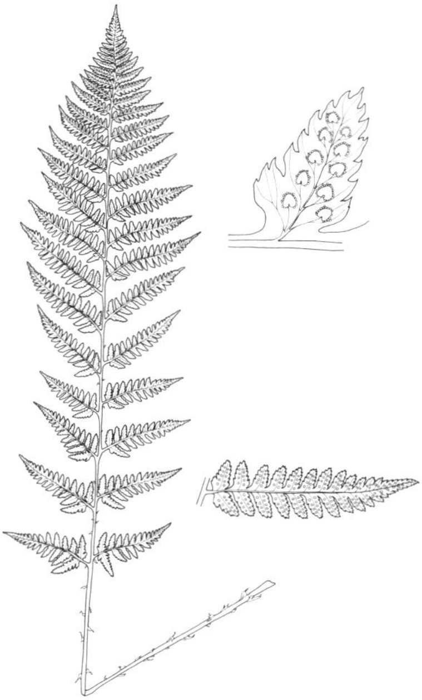

Distribution in New Jersey: Scattered in the northern counties; older records from Middlesex, Monmouth, and Burlington counties.

Comments: This hybrid is much less common in New Jersey than *D.* × *boottii,* although the parents frequently grow together. Usually only one individual is found in a locality. *D.* × *uliginosa* can be distinguished from *D.* × *boottii* by the lack of glandular hairs and, late in the summer, by the yellowing of fronds.

D. × *uliginosa* is tetraploid, with 41 pairs and 82 single chromosomes. The pairs indicate the relationship between the parents, but the ancestor in common between them is the unknown "Dryopteris semicristata."

Dryopteris carthusiana × *intermedia* *D.* × *triploidea* Wherry

Triploid Hybrid Wood Fern

Description: Fronds tripinnate, subevergreen (green well into winter but eventually turning brown); indusia and pinnule axes glandular.

Habitat: Swamps, damp woods, and thickets; conifer plantations; and roadside banks or stone walls.

Range: NS to MN, south to NC and KY.

Distribution in New Jersey: Common; recorded from 13 counties, mostly in the northern part of the state.

Comments: This is the most abundant fern hybrid in New Jersey, but it is difficult to distinguish from the parents. Where *D. intermedia* and *D. carthusiana* occur together, the hybrid can almost always be found, sometimes more abundantly than either of the parents. Plants of the hybrid are often the largest in the population, and those with the aspect of *D. carthusiana* and the glandular hairs of *D. intermedia* are probably hybrids. The aborted spores are the only certain means to distinguish *D.* × *triploidea* from the parents.

D. × *triploidea* is triploid, with 42 pairs and 42 single chromosomes. As with other hybrids, the chromosome pairs indicate the relationship between the parents.

Dryopteis carthusiana × *intermedia*

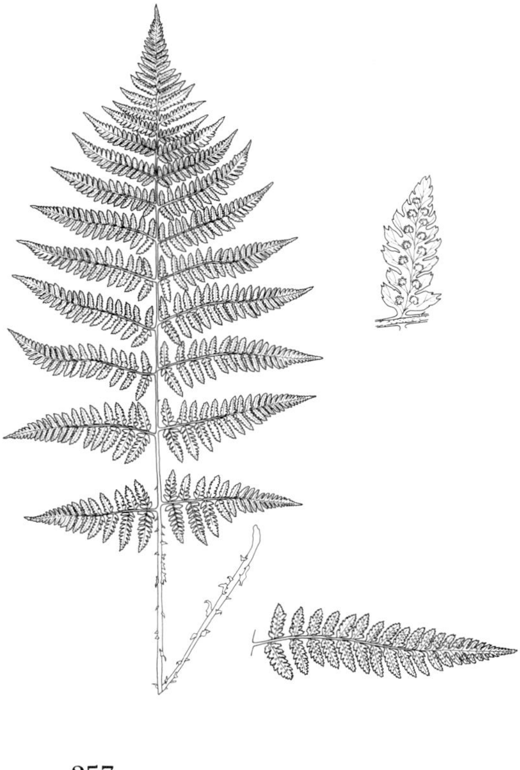

Dryopteris carthusiana × marginalis
D. × pittsfordensis Slosson

This hybrid has been recorded three times from New Jersey: once each from Sussex, Middlesex, and Monmouth counties (the Monmouth record may be a cultivated plant according to Montgomery, 1976). The most recent collection was in 1945. Both parents are abundant in northern New Jersey, but occur in different habitats. *D. × pittsfordensis* is rare throughout its wide range (Montgomery 1982a).

Dryopteris clintoniana × cristata

Description: Fronds pinnate-pinnatifid, fertile ones deciduous, usually with a few small evergreen vegetative fronds; blades lanceolate, much longer than wide; basal pinnae triangular; scales at base of stipe brown, nearly uniform in color.

Habitat: Swamps.

Range: CT, NY, ON, and MI, south to NJ and PA.

Distribution in New Jersey: Uncommon; Sussex, Passaic, Warren, and Morris counties.

Comments: *D. clintoniana × cristata* is usually overlooked, since the parents are very similar morphologically and in habitat requirements. The hybrid can be detected by examining spores of suspiciously intermediate plants (Montgomery 1982a).

This hybrid is pentaploid, with approximately 82 pairs and 41 single chromosomes. The pairs originate from the identity of two sets of chromosomes of *D. clintoniana* with *D. cristata,* indicating that *D. cristata* is one of the parents of *D. clintoniana; D. goldiana* is the other parent and the source of the single chromosomes.

Dryopteris clintoniana × cristata

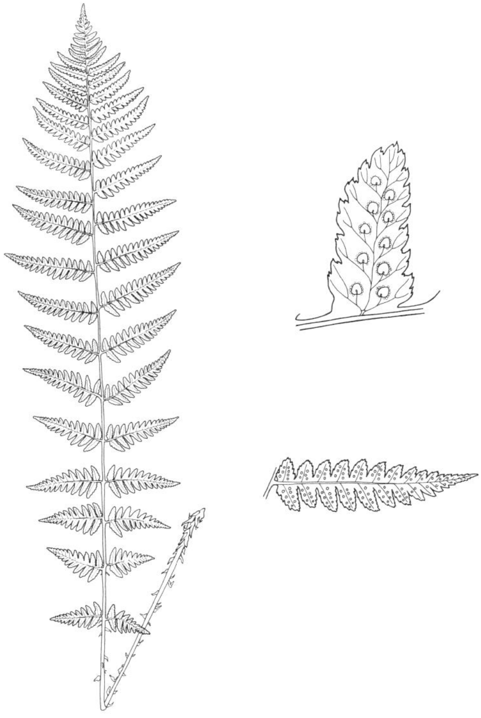

Dryopteris clintoniana × goldiana

Description: Fronds pinnate-pinnatifid to bipinnate at the base, deciduous; blades ovate, two to three times longer than wide; scales at base of stipe dark brown with paler border; sori closer to midrib than margin of the segments.

Habitat: Hardwood swamps or edges of swamps.

Range: NY, NJ, PA, ON, MI.

Distribution in New Jersey: Uncommon; Sussex, Warren, Passaic, and Essex counties.

Comments: This is one of the largest and most spectacular *Dryopteris* hybrids. It may be present at the edge of a swamp if *D. goldiana* grows on the slopes and *D. clintoniana* grows in the swamp.

This hybrid is tetraploid, with 41 pairs and 82 single chromosomes.

Dryopteris clintoniana × goldiana

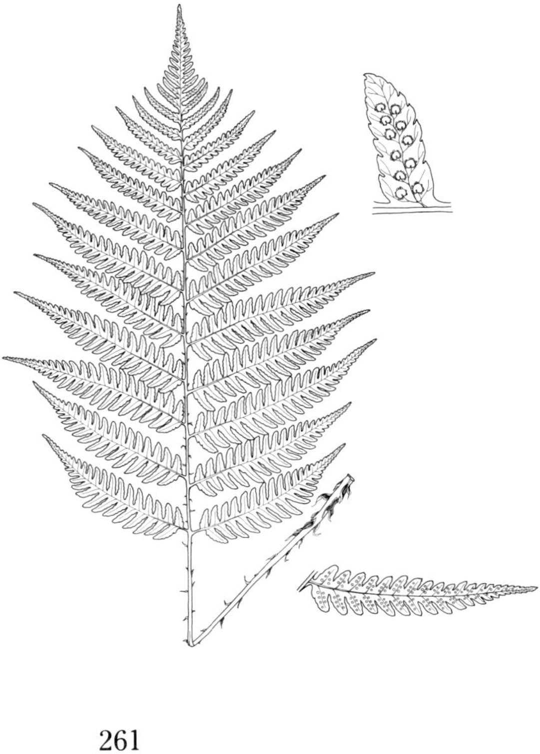

Dryopteris clintoniana × *intermedia*
D. × *dowellii* (Farw.) Wherry

Description: Fronds bipinnate-serrate to bipinnate-pinnatifid at the base, subevergreen; basal pinnae triangular; indusia and pinnule axes with glandular hairs.

Habitat: Swampy woods.

Range: NH to MI, south to NJ and PA.

Distribution in New Jersey: Several records for the northwestern counties, and an older record for Middlesex County.

Comments: *Dryopteris* × *dowellii* closely resembles *D.* × *boottii*, differing only in larger size and somewhat broader blades. The chromosome number is also different: *D.* × *dowellii* has 164 single chromosomes. Montgomery (1975) reported two chromosome counts from New Jersey.

This hybrid was first described by Philip Dowell from Lake Macopin, Passaic County, New Jersey, in 1908, and named for him by Farwell. Philip Dowell was one of the founders and the first editor of the *American Fern Journal*. He collected many *Dryopteris* species and hybrids in northern New Jersey and Staten Island, NY, correctly identifying many of them at a time when hybrids were not properly understood.

Dryopteris clintoniana × intermedia

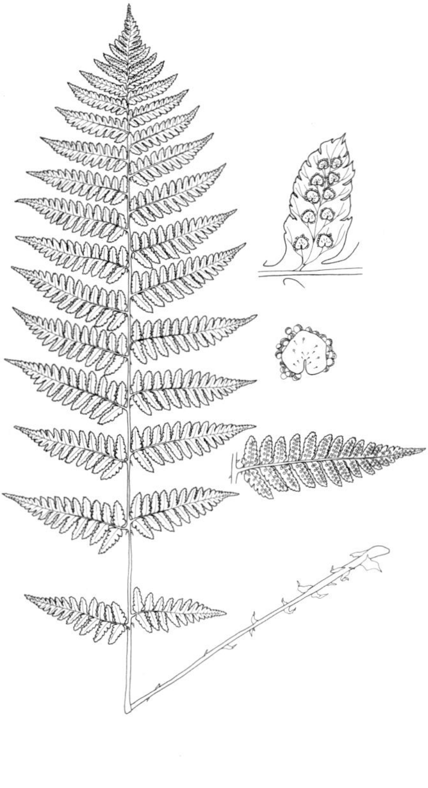

Dryopteris clintoniana × marginalis

Plants of this hybrid resemble *D. clintoniana* in blade shape and cutting, but the sori are nearer the margin than the midrib of the segments and the spores are aborted. There are five records from New Jersey: two from Sussex County, and one each from Essex, Passaic, and Warren counties.

Dryopteris cristata × goldiana

There is no certain record of this hybrid from New Jersey, but a specimen from Lodi, Bergen County, at the Staten Island Museum Herbarium might represent this combination. Individuals would be identical with *D. clintoniana,* but triploid with aborted spores.

Dryopteris cristata × intermedia
D. × boottii (Tuckerm.) Underw.

Boott's Hybrid Wood Fern

Description: Fronds evergreen, bipinnate at least at the base; basal pinnae triangular; indusia and pinnule axes with minute glandular hairs.

Habitat: Swamps, wet thickets, and moist woods.

Range: NF to WI, south to NJ, PA, WV, and TN.

Distribution in New Jersey: Common in the northern counties and a few scattered stations in central and southern New Jersey.

Comments: This is the best known of *Dryopteris* hybrids. It combines the characteristics of the parents, making it a good example of the intermediate morphology of a hybrid. This can be studied if a frond of the hybrid is placed with one of the parents on either side. *D.* × *boottii* is nearly always present when the parents occur together, and several individuals often can be found in larger populations.

D. × *boottii* is triploid, with 123 single chromosomes.

Dryopteris cristata × intermedia

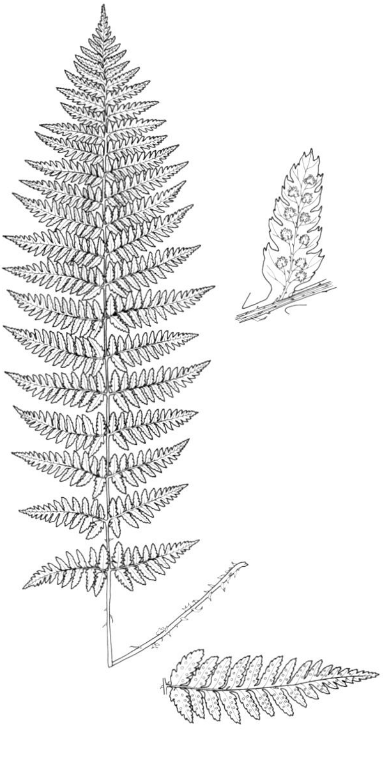

Dryopteris cristata × *marginalis*
D. × *slossonae* Wherry

Description: Fronds pinnate-pinnatifid, subevergreen, often somewhat irregular in construction, with misshaped or missing pinnae at the apex; blades lanceolate, 10–20 cm wide; lowest pinnae triangular; scales at base of stipes tan; sori nearer to margin than midrib of the segments.

Habitat: Swampy woods, stream banks, and moist wooded slopes.

Range: NB and ME to WI, south to VA and OH.

Distribution in New Jersey: Scattered over the northern counties, with older records for three central counties.

Comments: This hybrid is fairly easy to recognize with the combination of parental characteristics. Although the parents grow in quite different habitats, they may occur together where ridges or rock outcrops border swamps. The hybrid usually can be found if both parents are present.

D. × *slossonae* is triploid, with 123 single chromosomes. Counts were reported from New Jersey by Montgomery (1976). Sometimes the cutting of the frond is irregular, as if the plant were diseased.

Dryopteris cristata × *marginalis*

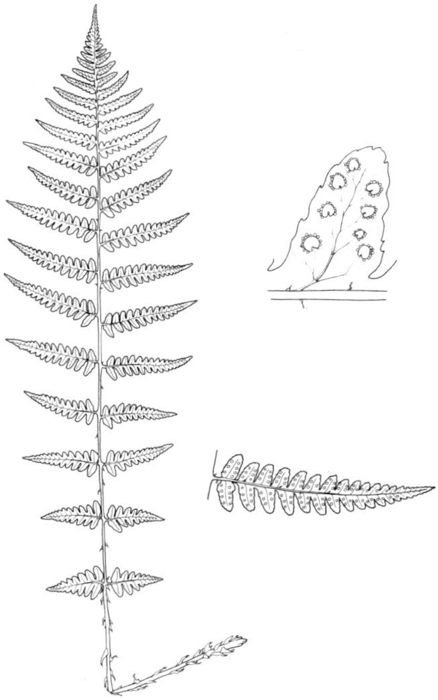

Dryopteris goldiana × intermedia

This striking hybrid was found by Dowell at West Englewood Station in 1916. There are no other records for the state. Although the parents have widely overlapping geographic ranges, the hybrid is rare throughout this range (Wagner 1971).

Dryopteris goldiana × marginalis *D. × neowherryi* Wagner

Description: Fronds pinnate-pinnatifid to bipinnate at the base, subevergreen, broadly ovate, often very large (up to 1 m long); pinnae ovate, often lobed; scales at base of stipes glossy brown with a paler border; sori closer to the margin than the midrib of the segments, or in the sinus of segment lobes.

Habitat: Rich, rocky woods.

Range: VT, NY, ON, south to w VA and w NC, west to AR and IL.

Distribution in New Jersey: Scattered in the northern counties.

Comments: *D. × neowherryi* is usually found where the parents grow together; however, since *D. goldiana* is uncommon in New Jersey, the hybrid is also uncommon. The hybrid is diploid, with 82 single chromosomes. *D. × neowherryi* may be very large, and makes a fine garden plant; however, the fronds die back late in the autumn. Plants will spread slowly by the short-creeping rhizome.

Dryopteris intermedia × marginalis

Although the parents of this hybrid are found together on rocky hillsides and ravines in many places in northern New Jersey, *D. intermedia × marginalis* has been collected only twice: in Sussex County (1914) and in Union County (1973). Montgomery (1982a) indicated that this hybrid is rare throughout the geographic range of the parents.

Dryopteris goldiana × marginalis

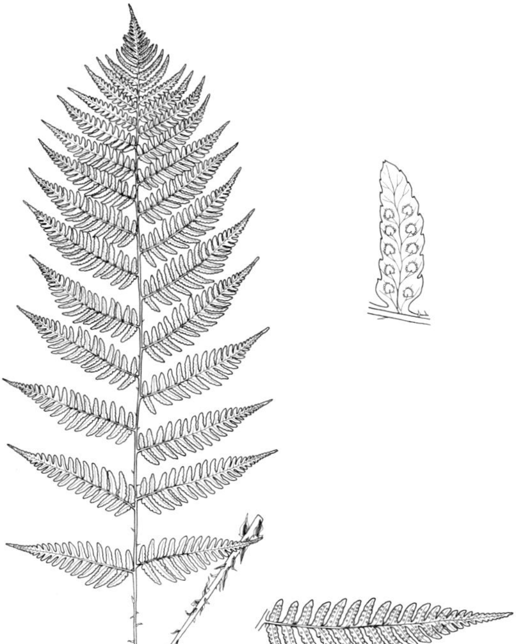

Woodwardia

Medium-sized to large woodland or swamp ferns. Rhizomes long-creeping; fronds deciduous, pinnatifid to pinnate-pinnatifid; monomorphic or dimorphic; sori elongate and arranged in a single row on each side of the midveins of the pinnae. Two species of this small genus occur in New Jersey.

Key to Species

1. Fronds dimorphic, the vegetative pinnate at base, pinnatifid above the base .... *W. areolata*
1. Fronds monomorphic, pinnate-pinnatifid except at the apex . *W. virginica*

Woodwardia areolata (L.) Moore

[*Lorinseria areolata* (L.) Presl]
Netted Chain Fern

Description: Fronds dimorphic, the vegetative pinnate at base, pinnatifid above the base, up to 60 cm tall, ovate, with areolate venation; the fertile fronds with linear pinnae, sori in rows along the pinna midrib.

Habitat: Acid wet woods and thickets; in the Pinelands at the edge of but rarely in Atlantic white cedar swamps.

Range: FL to TX, north, mostly on the Coastal Plain to se NH and NS; inland to NC, TN, and disjunct in MI, MO to OK.

Distribution in New Jersey: Common on the Coastal Plain and Piedmont wetlands; a few stations in bogs in the Highlands and Ridge and Valley.

Comments: Sterile plants of *W. areolata* may be confused with *Onoclea sensibilis,* since both species have creeping rhizomes, areolate venation, and occupy similar habitats. The lobes of the fronds are opposite in *Onoclea* but alternate in *W. areolata.*

This species is sometimes separated from *Woodwardia* into the segregate genus *Lorinseria,* because the fronds are dimorphic, but Lellinger (1985) indicated that there are often leaves that are partly fertile and partly vegetative. The related genus *Blechnum* also contains both monomorphic and dimorphic species.

The chromosome number of *W. areolata* is $2n = 70$.

Woodwardia areolata (L.) Moore

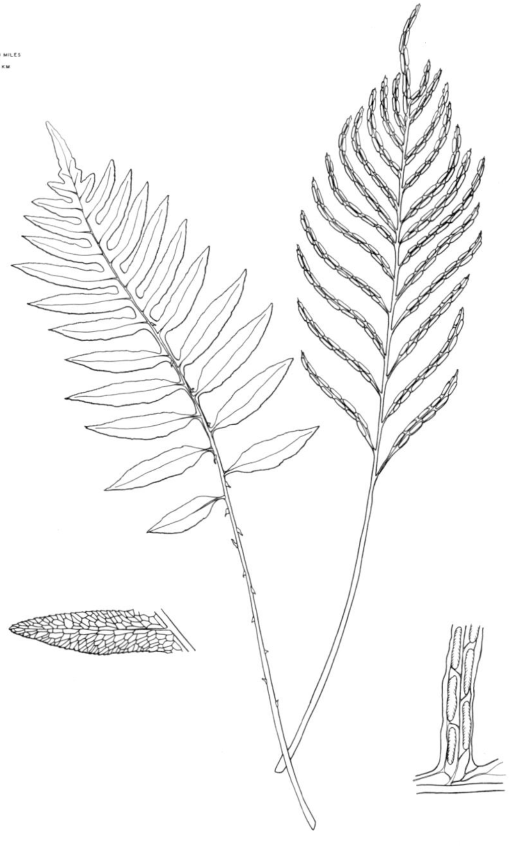

Woodwardia virginica (L.) J. E. Smith

[*Anchistea virginica* (L.) Presl]
Virginia Chain Fern

Description: Fronds all alike, up to 1 m tall; the stipes thick, nearly black at the base; blades pinnate-pinnatifid, ovate, veins areolate along the midrib of the pinnae; sori associated with these areolate veins.

Habitat: Bogs, open swamps, and wet thickets, always in acid soil or peat.

Range: FL to TX, north to NS, sw PQ, OH, and s MI; also Bermuda.

Distribution in New Jersey: Common to abundant in the southern part of the state, often invading abandoned cranberry bogs and in young Atlantic white cedar swamps in the Pinelands; scattered in the northern counties.

Comments: When not fertile, *W. virginica* may be confused with *Osmunda cinnamomea*. The two species often grow together, but they may be distinguished by the black scattered stipes and areolate venation of *Woodwardia* vs. the brown-to-tan clumped stipes and free veins of *Osmunda*. The tufts of hair in the pinna axils of cinnamon fern are another good distinguishing character.

The chromosome number of *W. virginica* is $2n = 70$.

Woodwardia virginica (L.) J. E. Smith

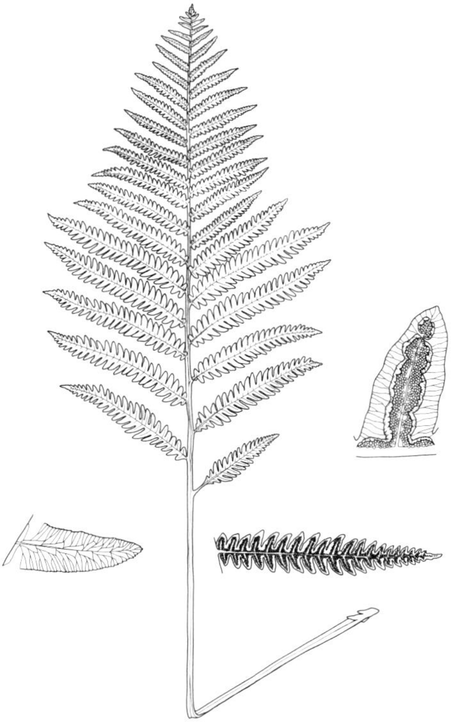

Marsilea

Plants of wet to damp habitats. Rhizomes long-creeping in mud; fronds with a long stipe and four pinnae arranged like a four-leaf clover; spores of two kinds, borne in hard-walled, globular structures (sporocarps) near the base of the stipes.

Marsilea contains about 60 species, widely distributed; one species has been introduced into New Jersey.

Marsilea quadrifolia L.

European Water-clover

Description: Fronds to 25 cm long when growing in water; much shorter when growing on mud, the four pinnae wedge-shaped.

Habitat: Ponds, slow-moving rivers, and marshes.

Range: Native to Europe; introduced to New England and ON to IA, KS, and KY.

Distribution in New Jersey: One station in Sussex County, and an old record from cultivated material in Morris County.

Comments: *M. quadrifolia* has been present in a pond in Sussex County for at least 40 years. The first introduction in America occurred in CT, in 1860, and it has become widespread, but not common, in the eastern United States (Johnson 1986).

In New Jersey, sporocarps can be found in early autumn. They are more likely to be found where plants are growing on mud, rather than in deeper water. In some parts of its range, water-clover plants grow in water up to 60 cm deep, with the rhizome rooted in soil at the bottom and the four pinnae floating on the water surface.

The chromosome number of *M. quadrifolia* is $2n = 40$.

Marsilea quadrifolia L.

Azolla

Plants floating on water. Rhizomes very fragile, readily separating into small fragments; fronds minute, scale-like, overlapping; spores of two kinds, borne in globular modified segments (sporocarps).

Azolla contains about six species (Lellinger 1985), widely distributed in both Old and New Worlds. One species occurs in the eastern United States, and two others in the western states.

Azolla caroliniana Willd.

Mosquito Fern

Description: Plants 0.5–1.0 cm long, floating on water.

Habitat: Canals, quiet water in streams, and ponds.

Range: FL to LA, north to NC, and locally to MA, NY, OH, and WI.

Distribution in New Jersey: Five records in as many counties.

Comments: *Azolla* is introduced into New Jersey from time to time, probably by waterfowl. In favorable years it forms huge colonies on canals, turning bright red in the autumn (Benedict 1923). T. C. Nelson collected it on the Raritan Canal near New Brunswick in 1949, and it persisted for at least two years (Cohn and Renlund 1953) but eventually disappeared. There is no record of it persisting longer than this in the state.

The only way to precisely identify the species of *Azolla* is to study the sporocarps. Since none of the herbarium material from New Jersey is fertile, we can not be certain of the species. Several species are grown in aquaria and outdoor pools, from which they might be introduced to natural sites.

Azolla caroliniana Willd.

Checklist of the Ferns and Fern-Allies of New Jersey

Taxon	*Common name*
Lycopodiaceae	
Lycopodium alopecuroides	Foxtail clubmoss
Lycopodium annotinum	Bristly clubmoss
Lycopodium appressum	Appressed bog clubmoss
Lycopodium carolinianum	Carolina clubmoss
Lycopodium clavatum var. *clavatum*	Staghorn clubmoss
Lycopodium dendroideum	Tree clubmoss
Lycopodium digitatum	Common running-pine
Lycopodium inundatum	Bog clubmoss
Lycopodium lucidulum	Shining firmoss
Lycopodium obscurum var. *obscurum*	Tree clubmoss
Lycopodium obscurum var. *isophyllum*	Tree clubmoss
Lycopodium tristachyum	Ground-cedar
Lycopodium × *copelandii*	Hybrid bog clubmoss
Lycopodium × *habereri*	Hybrid running-pine
Lycopodium alopecuroides × *inundatum*	Hybrid bog clubmoss
Lycopodium appressum × *inundatum*	Hybrid bog clubmoss
Selaginellaceae	
Selaginella apoda	Meadow spikemoss
Selaginella rupestris	Rock spikemoss
Isoetaceae	
Isoetes echinospora	Spiny-spored quillwort
Isoetes engelmannii	Engelmann's quillwort
Isoetes lacustris	Lake quillwort
Isoetes melanopoda	Black-footed quillwort
Isoetes riparia	Shore quillwort
Isoetes tuckermanii	Tuckerman's quillwort
Isoetes × *dodgei*	Dodge's quillwort
Isoetes × *eatonii*	Eaton's quillwort
Isoetes engelmannii × *lacustris*	Hybrid quillwort
Isoetes engelmannii × *riparia*	Hybrid quillwort
Equisetaceae	
Equisetum arvense	Field horsetail
Equisetum fluviatile	Water horsetail
Equisetum hyemale var. *affine*	Scouring-rush
Equisetum pratense	Meadow horsetail
Equisetum sylvaticum	Woodland horsetail
Equisetum variegatum	Variegated scouring-rush
Equisetum × *ferrissii*	Intermediate scouring-rush
Equisetum × *litorale*	Shore horsetail
Equisetum × *trachyodon*	Hybrid scouring-rush

Checklist (*continued*)

Taxon	*Common name*
Ophioglossaceae	
Botrychium dissectum	Cut-leaved grape fern
Botrychium lanceolatum subsp. *angustisegmentum*	Lance-leaved moonwort
Botrychium matricariifolium	Daisy-leaved moonwort
Botrychium multifidum	Leathery grape fern
Botrychium oneidense	Blunt-lobed grape fern
Botrychium simplex	Least moonwort
Botrychium virginianum	Rattlesnake fern
Ophioglossum vulgatum	Southern adder's-tongue
Ophioglossum pusillum	Northern adder's-tongue
Osmundaceae	
Osmunda cinnamomea	Cinnamon fern
Osmunda claytoniana	Interrupted fern
Osmunda regalis	Royal fern
Schizaeaceae	
Schizaea pusilla	Curly-grass fern
Lygodium palmatum	Climbing fern
Adiantaceae	
Adiantum pedatum	Maidenhair fern
Cryptogramma stelleri	Slender rock-brake
Cheilanthes lanosa	Hairy lip fern
Pellaea atropurpurea	Purple cliff-brake
Pellaea glabella	Smooth cliff-brake
Hymenophyllaceae	
Trichomanes sp.	Appalachian gametophyte
Polypodiaceae	
Polypodium virginianum	Common polypody
Dennstaedtiaceae	
Dennstaedtia punctilobula	Hay-scented fern
Pteridium aquilinum var. *latiusculum*	Northeastern bracken
Pteridium aquilinum var. *pseudocaudatum*	Southeastern bracken
Thelypteridaceae	
Thelypteris hexagonoptera	Southern beech fern
Thelypteris noveboracensis	New York fern
Thelypteris palustris var. *pubescens*	Marsh fern
Thelypteris phegopteris	Northern beech fern
Thelypteris simulata	Massachusetts fern
Aspleniaceae	
Asplenium bradleyi	Bradley's spleenwort
Asplenium montanum	Mountain spleenwort
Asplenium pinnatifidum	Lobed spleenwort
Asplenium platyneuron	Ebony spleenwort
Asplenium rhizophyllum	Walking fern
Asplenium ruta-muraria	Wall-rue
Asplenium trichomanes	Maidenhair spleenwort
Asplenium × *ebenoides*	Scott's spleenwort
Asplenium × *trudellii*	Trudell's spleenwort

Checklist (*continued*)

Taxon	*Common name*
Asplenium × *wherryi*	Wherry's spleenwort
Matteuccia struthiopteris var. *pensylvanica*	Ostrich fern
Onoclea sensibilis	Sensitive fern
Athyrium filix-femina var. *angustum*	Northern lady fern
Athyrium filix-femina var. *asplenioides*	Southern lady fern
Athyrium pycnocarpon	Glade fern
Athyrium thelypterioides	Silvery spleenwort
Gymnocarpium dryopteris var. *dryopteris*	Oak fern
Cystopteris bulbifera	Bulblet fern
Cystopteris protrusa	Lowland fragile fern
Cystopteris tenuis	Mackay's fragile fern
Woodsia ilvensis	Rusty woodsia
Woodsia obtusa	Blunt-lobed woodsia
Polystichum acrostichoides	Christmas fern
Dryopteris carthusiana	Spinulose wood fern
Dryopteris celsa	Log Fern
Dryopteris clintoniana	Clinton's wood fern
Dryopteris cristata	Crested wood fern
Dryopteris goldiana	Goldie's wood fern
Dryopteris intermedia	Evergreen wood fern
Dryopteris marginalis	Marginal wood fern
Dryopteris × *benedictii*	Benedict's wood fern
Dryopteris × *boottii*	Boott's wood fern
Dryopteris × *dowellii*	Dowell's wood fern
Dryopteris × *leedsii*	
Dryopteris × *neowherryi*	Wherry's wood fern
Dryopteris × *pittsfordensis*	
Dryopteris × *separabilis*	
Dryopteris × *slossonae*	Slosson's wood fern
Dryopteris × *triploidea*	Triploid wood fern
Dryopteris × *uliginosa*	
Dryopteris carthusiana × *celsa*	
Dryopteris celsa × *cristata*	
Dryopteris celsa × *goldiana*	
Dryopteris clintoniana × *cristata*	
Dryopteris clintoniana × *goldiana*	
Dryopteris clintoniana × *marginalis*	
Dryopteris goldiana × *intermedia*	
Dryopteris intermedia × *marginalis*	
Blechnaceae	
Woodwardia areolata	Netted chain fern
Woodwardia virginica	Virginia chain fern
Marsileaceae	
Marsilea quadrifolia	European water-clover
Azollaceae	
Azolla caroliniana	Mosquito fern

References

Atkinson, L. R. 1965. *Schizaea pusilla*—perennial gametophyte? Amer. Fern J. 55:81–83.

Balick, M. J., and J. M. Beitel. 1989. *Lycopodium* spores used in condom manufacture: associated health hazards. Econ. Bot. 43:373–377.

Barrington, D. S., C. H. Haufler, and C. R. Werth. 1989. Hybridization, reticulation, and species concepts in the ferns. Amer. Fern. J. 79:55–64.

Barton, W. P. C. 1818. *Compendium Florae Philadelphicae,* 2 vols. Carey & Son, Philadelphia.

Beitel, J. M. 1979. Clubmosses (*Lycopodium*) in North America. Fiddlehead Forum 6(5):1–8.

———, and F. S. Wagner. 1982. The chromosomes of *Lycopodium lucidulum.* Amer. Fern J. 72:33–35.

Benedict, R. C. 1923. The mosquito fern. Amer. Fern J. 13:48–52.

———. 1936. Field trip of September 27, 1936, at Branchville, N.J. Amer Fern J. 26:142–143.

Berkeley, E., and D. S. Berkeley. 1982. *The Life and Travels of John Bartram: From Lake Ontario to the River St. John.* University Presses of Florida, Tallahassee.

Boivin, B. 1950. The problem of generic segregates in the form-genus *Lycopodium.* Amer. Fern J. 40:32–41.

Braun, A. 1846. Über die nordamericanischen *Isoetes*-Arten. Flora (Regensburg Bot. Zeit.) 29:177–180.

Braun, E. L. 1950. *Deciduous Forests of Eastern North America.* Blakiston, Philadelphia.

Britton, D. M. 1976. The distribution of *Dryopteris spinulosa* and its relatives in eastern Canada. Amer. Fern J. 66:69–74.

Britton, E. G. 1897. A revision of the North American species of *Ophioglossum.* Bull. Torrey Bot. Club 24:545–559.

———, and A. Taylor. 1901. Life history of *Schizaea pusilla.* Bull. Torrey Bot. Club 28:1–19.

Britton, N. L. 1881. *A Preliminary Catalogue of the Flora of New Jersey.* NJ Geological Survey, New Brunswick, NJ.

———. 1889. *Catalogue of Plants Found in New Jersey: Final Report of the State Geologist,* Vol. II, Pt. 1. NJ Geological Survey Rept. Murphy Publ., Trenton, NJ.

Brown. H. H. 1945. *Schizaea pusilla* from Ontario, Canada. Amer. Fern J. 77:64–65.

Brown, O. H., and E. T. Wherry. 1970. Check list of the vascular flora of Cape May County, New Jersey. Bartonia 40:1–18.

Bruce, J. G. 1975. Systematics and morphology of subgenus *Lepidotis* of the genus *Lycopodium* (Lycopodiaceae). Ph.D. dissertation, University of Michigan, Ann Arbor.

———. 1976. Comparative studies in the biology of *Lycopodium carolinianum.* Amer. Fern J. 66:125–137.

———, and A. M. Evans. 1979. The correct name for a widespread hybrid *Lycopodium.* Taxon 28:413–414.

Buchholz, K., and R. E. Good. 1982. *Compendium of New Jersey Pine Barren Literature* (rev. 1983). Center for Coastal and Environmental Studies, Rutgers, The State University, New Brunswick, NJ.

Chrysler, M. A., and J. L. Edwards. 1947. *The Ferns of New Jersey.* Rutgers University Press, New Brunswick, NJ.

Clausen, R. T. 1938. A monograph of the Ophioglossaceae. Mem. Torrey Bot. Club. 19(2): 1–177.

Cohn, J., and R. N. Renlund. 1953. Notes on *Azolla caroliniana.* Amer. Fern J. 43:7–11.

Collins, B. R., and E. W. B. Russell, eds. 1988. *Protecting the New Jersey Pinelands.* Rutgers University Press, New Brunswick, NJ.

Crabbe, J. A., A. C. Jermy, and J. T. Mickel. 1975. A new generic sequence for the pteridophyte herbarium. Fern Gaz. 11:141–162.

Crawford, J. 1908. Some sand dune plants from Longport, New Jersey. Bartonia 1:18–19.

de la Pylaie, B. 1826. Sur l'île de Terre-Neuve et quelques îles voisines. Mem. Soc. Linn. Paris. 4:424.

Dole, W. H. 1948. The passing of a New Jersey fern station. Amer. Fern J. 38:19–20.

Dowell, P. 1908. New ferns described as hybrids in the genus *Dryopteris*. Bull. Torr. Bot. Club 35:137–140.

Dunbar, L. 1989. *Ferns of the Coastal Plain*. University of South Carolina Press, Columbia.

Eaton, D. C. 1879. *The Ferns of North America: with Synonymy and Geographical Distribution of the Ferns of the United States of America and the British North American Possessions*, 2 vols. Cassino Publ., Salem, MA.

Engelmann, G. 1882. The genus *Isoetes* in North America. Trans. St. Louis Acad. Sci. 4:358–390.

Ewan, J. 1952. Frederick Pursh, 1774–1820, and his botanical associates. Proc. Amer. Philos. Soc. 96:599–628.

———, ed. 1974. *Classica Botanica Americana*, Vol. 3. Hafner Press, New York. (Facsimile of 1803 edition of *Flora Boreali Americana* by André Michaux.)

Fairbrothers, D. E. 1989. Plants in jeopardy: past, present, future. Pp. 5–12 *in* E. Karlin, ed., *New Jersey's Rare and Endangered Plants and Animals*. Institute for Environmental Studies, Ramapo College, Mahwah, NJ.

———, and M. Y. Hough. 1973. Rare or endangered vascular plants of New Jersey. Science Notes, No. 14. NJ State Museum, Trenton, NJ.

Farrar, D. R., J. C. Parks, and B. W. McAlpin. 1983. The fern genera *Vittaria* and *Trichomanes* in the northeastern United States. Rhodora 85:83–92.

Ferguson, G. 1959. *Signs and Symbols in Christian Art*. Oxford University Press, New York.

Fernald, M. L. 1950. *Gray's Manual of Botany*, 8th ed. American Book Co., New York.

Forman, R. T. T., ed. 1979. *Pine Barrens: Ecosystem and Landscape*. Academic Press, New York.

Fraser-Jenkins, C. R. 1980. Nomenclatural notes on *Dryopteris*. 4. Taxon 29:607–612.

Goldsworthy, W. J., Jr., and M. D. Shulman. 1985. Growing season trends for New Jersey. Bull. N.J. Acad. Sci. 30:17–21.

Guiragossian, H. A., and R. E. Koning. 1986. Induction of spore germination in *Schizaea pusilla* (Schizaeaceae). Amer. J. Bot. 73:1588–1594.

Harshberger, J. W. 1916. *The Vegetation of the New Jersey Pine Barrens* (Dover reprint, 1970). Dover Publ., New York.

Haufler, C. H. 1985. Pteridophyte evolutionary biology: the electrophoretic approach. Proc. Roy. Soc. Edinburgh 86:315–323.

———, and M. D. Windham. 1991. New species of North American *Cystopteris* and *Polypodium*, with comments on their reticulate relationships. Amer. Fern J. 81:7–23.

Hauke, R. L. 1963. A taxonomic monograph of the genus *Equisetum* subgenus *Hippochaete*. Nova Hedwigia 8:1–123.

———. 1966. A systematic study of *Equisetum arvense*. Nova Hedwigia 13:81–109.

———. 1978. A taxonomic monograph of *Equisetum* subgenus *Equisetum*. Nova Hedwigia 30:385–455.

Hickey, R. J. 1977. The *Lycopodium obscurum* complex in North America. Amer. Fern J. 67:45–48.

———. 1986. *Isoetes* megaspore surface morphology: nomenclature, variation, and systematic importance. Amer. Fern J. 76:1–16.

———, and J. M. Beitel. 1979. A name change for *Lycopodium flabelliforme*. Rhodora 81:137–140.

Hirst, F. 1990. Three new taxa for the Delmarva Peninsula. Bartonia 56:70–71.

Holttum, R. E. 1949. The classification of ferns. Biol. Rev. 24:267–296.

———. 1971. Studies in the family Thelypteridaceae III. A new system of genera in the Old World. Blumea 19:17–52.
———. 1983. The family Thelypteridaceae in Europe. Acta Bot. Malacitana 8:47–58.
Holub, J. 1985. Transfers of *Lycopodium* species to *Huperzia:* with a note on generic classification in Huperziaceae. Folia Geobot. Phytotax. 20:67–80.
House, H. D. 1904. Some rare ferns of central New Jersey. Fern Bull. 12:80–82.
Howe, M. A. 1934. Nathaniel Lord Britton. J. N.Y. Bot. Gard. 35:169–180.
Jarrett, W. F. H. 1982. Bracken and cancer. Proc. Roy. Soc. Edinburgh 81B:79–83.
Jobes, G. 1962. *Dictionary of Mythology, Folklore, and Symbols,* Pts. 1 & 2. Scarecrow Press, New York.
Johnson, D. M. 1986. Systematics of the New World species of *Marsilea* (Marsileaceae). Syst. Bot. Monogr. 11:1–87.
Juel, H. O., and J. W. Harshberger. 1929. New light on the collection of North American plants made by Peter Kalm. Proc. Acad. Nat. Sci. Phila. 81:297–303.
Kalm, P. 1972. *Travels into North America* (reprint of English translation by J. R. Forster, 1770–1771). The Imprint Society, Barre, MA.
Kato, M. 1977. Classification of *Athyrium* and allied genera of Japan. Bot. Mag. Tokyo 90:23–40.
———, and D. Darnaedi. 1988. Taxonomic and phytogeographic relationships of *Diplazium flavoviride, D. pycnocarpon,* and *Diplaziopsis.* Amer. Fern J. 78:77–85.
———, and K. Iwatsuki. 1983. Phytogeographic relationships of pteridophytes between temperate North America and Japan. Ann. Missouri Bot. Gard. 70:724–733.
Kirkwood, R. C., and A. Hinshalwood. 1985. Recent studies on the biology of bracken and the mode of action of herbicides used in its control. Proc. Roy. Soc. Edinburgh 84B:179–185.
Knieskern, P. D. 1857. *Catalogue of Plants Growing Without Cultivation in the Counties of Monmouth and Ocean, New Jersey.* True American Office, Trenton, NJ.
Kott, L. S. 1981. *Isoetes acadiensis,* a new species from eastern North America. Can. J. Bot. 59:2592–2594.
———, and D. M. Britton. 1980. Chromosome numbers for *Isoetes* in northeastern North America. Can. J. Bot. 58:980–984.
———, and ———. 1982. A comparative study of sporophyte morphology of the three cytotypes of *Polypodium virginianum* in Ontario. Can. J. Bot. 60:1360–1370.
———, and ———. 1983. Spore morphology and taxonomy of *Isoetes* in northeastern North America. Can. J. Bot. 61:3140–3163.
LaWall, C. H. 1910. Botanical notes from Longport, N.J. Bartonia 3:12–21.
Lellinger, D. B. 1985. *A Field Manual of the Ferns and Fern Allies of the United States and Canada.* Smithsonian Institution Press, Washington, DC.
Liew, F. S. 1972. Numerical taxonomic studies on North American lady ferns and their allies. Taiwania 17:190–221.
Long, B. 1922. Occurrence of *Botrychium matricariaefolium* in New Jersey. Amer. Fern J. 12:1–9.
Löve, A., D. Löve, and R. E. G. Pichi Sermolli. 1977. *Cytotaxonomical Atlas of the Pteridophyta.* Cramer, Vaduz, Liechtenstein.
McCormick, J. 1979. The vegetation of the New Jersey Pine Barrens. Pp. 229–243 *in* R. T. T. Forman, ed., *Pine Barrens: Ecosystem and Landscape.* Academic Press, New York.
McPhee, J. 1968. *The Pine Barrens.* Farrar, Straus, & Giroux, New York.
Manton, I. 1950. *Problems of Cytology and Evolution in the Pteriodophyta.* Cambridge University Press, Cambridge.
Mickel, J. T. 1979. *How to Know the Ferns and Fern Allies.* Brown, Dubuque, IA.

Moldenke, H. N. 1960. Curlygrass in New York State. Rhodora 62:294.

Montgomery, J. D. 1975. *Dryopteris celsa* and *D. clintoniana* in New Jersey. Amer. Fern J. 65:65–69.

———. 1976. The distribution and abundance of *Dryopteris* in New Jersey. Amer. Fern J. 76:53–59.

———. 1978. A vegetative key to the ferns of the northeastern United States. Fiddlehead Forum 5(3):3–4.

———. 1981. *Equisetum variegatum* and *E.* × *trachyodon* in New Jersey. Amer. Fern J. 71:1–2.

———. 1982a. *Dryopteris* in North America. Part II. The hybrids. Fiddlehead Forum 9:23–30.

———. 1982b. Habitat preservation and rare pteridophytes in New Jersey. Pp. 294–301 *in* W. J. Cromartie, ed., *New Jersey's Endangered and Threatened Plants and Animals.* Center for Environmental Research, Stockton State College, Pomona, NJ.

———. 1989. Rare and endangered pteridophytes in New Jersey. Pp. 17–21 *in* E. Karlin, ed., *New Jersey's Rare and Endangered Plants and Animals.* Institute for Environmental Studies, Ramapo College, Mahwah, NJ.

———, and D. E. Fairbrothers. 1963. A floristic comparison of the vascular plants of two sphagnous wet lands in New Jersey. Bull. Torrey Bot. Club 90:87–99.

———, and ———. 1985. Geographical distribution of New Jersey's pteridophytes. Bartonia 51:52–57.

———, and E. M. Paulton. 1981. *Dryopteris* in North America. Fiddlehead Forum 8:25–31.

Moran, R. C. 1982. The *Asplenium trichomanes* complex in the United States and adjacent Canada. Amer. Fern J. 72:5–11.

Morton, C. V. 1963. The classification of *Thelypteris.* Amer. Fern J. 53:149–154.

———. 1967. Fern herbarium of André Michaux. Amer. Fern J. 57:166–182.

Muhlenberg, G. H. E. 1813. *Catalogues Plantarum Americae Septentrionalis huc usque Cognitarum Indigenarum et Cicurum.* Wm. Hamilton, Lancaster, PA.

———. 1818. *Catalogues Plantarum Americae Septentrionalis huc usque Cognitarum Indigenarum et Cicurum,* 2nd ed., corrected and enlarged. S. W. Conrad, Publ., Philadelphia.

Mulligan, G. A. and W. J. Cody. 1982. Chromosome numbers in Canadian *Phegopteris.* Can. J. Bot. 57:1815–1819.

Nuttall, T. 1818. *The Genera of North American Plants and a Catalogue of the Species to the Year 1817,* 2 vols. D. Heartt, Philadelphia.

Oinonen, E. 1967. Sporal regeneration of ground pine (*Lycopodium complanatum*) in southern Finland in the light of the size and the age of its clones. Acta For. Fenn. 83:76–85.

Øllgaard, B. 1979. Studies in the Lycopodiaceae. II. The branching patterns and intergeneric groups of *Lycopodium* sensu lato. Amer. Fern J. 69:49–61.

———. 1987. A revised classification of the Lycopodiaceae, s. lat. Opera Bot. 92:153–178.

Paris, C. A., and M. D. Windham. 1988. A biosystematic investigation of the *Adiantum pedatum* complex in eastern North America. Syst. Bot. 13:240–255.

Parsons, F. T. 1899. *How to Know the Ferns* (Dover reprint, 1961). Dover Publ., New York.

Parsons, M. R., and M. D. Shulman. 1978. Extreme value analysis of annual minimum temperatures in New Jersey. Bull. N.J. Acad. Sci. 23:43–49.

Peck, J. H. 1980. *Equisetum* × *litorale* in Illinois, Iowa, Minnesota, and Wisconsin. Amer. Fern J. 70:33–38.

Pichi Sermolli, R. E. G. 1973. Historical review of the higher classification of the Filicopsida. Pp. 11–40 & pl. 1–19 *in* A. C. Jermy, J. A. Crabbe, and B. A. Thomas, eds., *The Phylogeny and Classification of the Ferns.* Academic Press, London.

———. 1977. Tentamen pteridophytorum genera in taxonomicum ordinum redigendi. Webbia 31:313–512.

Pursh, F. T. 1814. *Florae Americae Septentrionalis,* 2 vols. White, Cochrane & Co., London.
Redfield, J. H. 1869. Note in Meeting of Academy of Natural Sciences of Philadelphia, 21 April 1869. Proc. Acad. Nat. Sci. Phila. 21:13.
———. 1876. Dr. Torrey's first trip to the Pines. Bull. Torrey Bot. Club 6:82–84.
Robichaud, B., and M. F. Buell. 1973. *Vegetation of New Jersey.* Rutgers University Press, New Brunswick, NJ.
Smith, A. R. 1971. Systematics of the neotropical species of *Thelypteris* section *Cyclosorus.* Univ. Calif. Publ. Bot. No. 59.
Snyder, D. B. 1984. New Jersey's threatened plant species. NJ Office of Natural Lands Management, Trenton, NJ.
———. 1985. Additions of New Jersey's flora. Bartonia 51:95–98.
———. 1988. *Special Plants of New Jersey.* NJ Natural Heritage Program, Trenton, NJ.
———, and V. E. Vivian. 1981. *Rare and Endangered Vascular Plant Species in New Jersey.* U.S. Fish and Wildlife Service, Washington, DC.
Stolze, R. M. 1987. *Schizaea pusilla* discovered in Peru. Amer. Fern J. 77:64–65.
Stone, W. 1911. The plants of southern New Jersey, with especial reference to the flora of the Pine Barrens. Ann. Rept. N.J. State Mus. (Trenton), 1910, pp. 25–828.
Taylor, N. 1915. *Flora of the Vicinity of New York.* Mem. N.Y. Bot. Gard. No. 5. Bronx, NY.
Taylor, W. C. 1984. *Arkansas Ferns and Fern Allies.* Milwaukee Public Museum, Milwaukee, WI.
———, and N. T. Luebke. 1986. Germinating spores and growing sporelings of aquatic *Isoetes.* Amer. Fern J. 76:21–24.
———, ———, and M. B. Smith. 1985. Speciation and hybridization in North American quillworts. Proc. Roy. Soc. Edinburgh 86B:259–263.
Tedrow, J. C. F. 1986. *Soils of New Jersey.* Krieger Publ., Malabar, FL.
Torrey, J., W. C. Eddy, and D. V. Knevels. 1819. *Catalogue of Plants Growing Spontaneously Within Thirty Miles of the City of New York.* Lyceum of Natural History of New York. Webster & Shinners, Albany, NY.
Tryon, A. F. 1957. A revision of the fern genus *Pellaea,* section *Pellaea.* Ann. Missouri Bot. Gard. 44:125–193.
———, and R. Tryon. 1973. *Thelypteris* in eastern North America. Amer. Fern J. 63:65–76.
Tryon, R. 1969. Pteridology. Pp. 97–102 *in* J. Ewan, ed., *A Short History of Botany in the United States.* Hafner Publ., New York.
———, and A. F. Tryon. 1982. *Ferns and Allied Plants, with Special Reference to Tropical America.* Springer-Verlag, New York.
Wagner, W. H., Jr. 1954. Reticulate evolution in the Appalachian Aspleniums. Evolution 7:103–118.
———. 1963. A biosystematic study of United States ferns—preliminary abstract. Amer. Fern J. 53:1–16.
———. 1971. Evolution of *Dryopteris* in relation to the Appalachians. *In* P. C. Holt, ed., *The Distributional History of the Biota of the Southern Appalachians,* Pt. II: *Flora.* Virg. Polytech. Inst. Res. Div. Monogr. 2:147–192.
———, J. M. Beitel, and R. C. Moran. 1989. *Lycopodium hickeyi:* a new species of North American clubmoss. Amer. Fern J. 79:119–121.
———, and T. Darling, Jr. 1957. Synthetic and wild *Asplenium gravesii.* Brittonia 7:57–63.
———, and F. S. Wagner. 1965. Rochester area log ferns (*Dryopteris celsa*) and their hybrids. Proc. Rochester Acad. Sci. 11:57–71.
———, and ———. 1966. Pteridophytes of the Mountain Lake area, Giles Co., Virginia: biosystematic studies 1964–65. Castanea 31:121–140.
———, and ———. 1983. Genus communities as a systematic tool in the study of New World *Botrychium* (Ophioglossaceae). Taxon 32:51–63.

———, ———, and D. J. Hagenah. 1969. The log fern, *Dryopteris celsa,* and its hybrids in Michigan—a preliminary report. Mich. Bot. 8:137–145.

———, ———, C. N. Miller, Jr., and D. H. Wagner. 1978. New observations on the royal fern hybrid *Osmunda* × *ruggii.* Rhodora 80:92–106.

Walker, S. 1955. Cytogenetic studies in the *Dryopteris spinulosa* complex. I. Watsonia 3:193–209.

———. 1961. Cytogenetic studies in the *Dryopteris spinulosa* complex. II. Amer. J. Bot. 48:607–614.

———. 1962. Further studies on the genus *Dryopteris:* the origin of *D. clintoniana, D. celsa,* and related taxa. Amer. J. Bot. 49:497–503.

Weatherby, U. F. 1952. The English names of North American ferns. Amer. Fern J. 42:134–150.

Wherry, E. T. 1935. Fern field notes, 1935. Amer. Fern J. 25:123–125.

———. 1961. *The Fern Guide.* Doubleday, Garden City, NY.

Wilce, J. H. 1965. Section *Complanata* of the genus *Lycopodium.* Nova Hedwigia 19:1–233, pl. I–XL.

———. 1972. Lycopod spores. I. General spore patterns and the generic segregation of *Lycopodium.* Amer. Fern J. 62:65–79.

Willis, O. R. 1874. *Catalogue of Plants Growing Without Cultivation in the State of New Jersey.* Schermerhorn Co., New York.

———. 1877. *Catalog of the Plants Growing Without Cultivation in the State of New Jersey,* rev. ed. Barnes, New York.

Wolfe, P. E. 1977. *The Geology and Landscapes of New Jersey.* Crane, Russak, New York.

Index